THE BIOTECH PRIMER

BioTech Primer Inc.

D1441732

THE
BIOTECH PRIMER

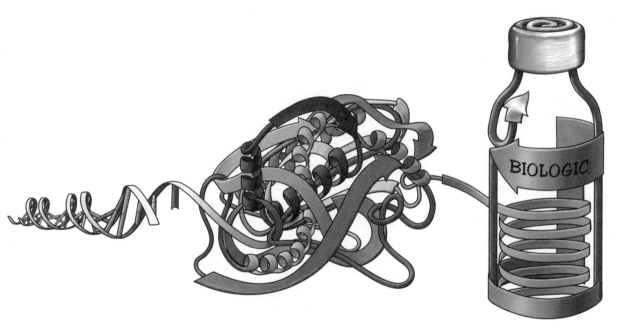

written by

biotech Primer

Industry Knowledge Delivered

This one image represents the science upon which the
biotech healthcare industry is built. If you don't understand it
now, you will after reading The Biotech Primer.

THE BIOTECH PRIMER

Published by:
BioTech Primer Inc.
1200 East Joppa Road
Towson, MD 21286

Email (for orders and customer service inquiries): info@BiotechPrimerInc.com
Visit our Home Page at www.BioTechPrimerInc.com
Copyright @ 2012 BioTech Primer Inc., Towson, Maryland, USA
Published by BioTech Primer Inc., Towson, Maryland, USA

ISBN 978-1477657256

About BioTech Primer

BioTech Primer Inc. educates decision makers impacting the biotechnology and pharmaceutical industries across the United States and around the world. With continuously updated materials and industry experienced instructors, BioTech Primer Inc. is uniquely positioned to provide the most current biotechnology instruction available. Clear, comprehensive, and compelling training gives participants a full understanding of the relevant science and technology inherent in the work they do. BioTech Primer Inc. products and services include courses, publications, and consultancy work.

Learn:

On-Site: Customized training at your facility. We develop and deliver courses to meet your needs.

Off-Site: Comprehensive list of courses delivered by industry experts in various locations throughout the USA.

Online: Online e-learning courses and on-demand webinars are available as a cost effective resource whenever needed.

Attend:

- BioImmersion
- BioBasics
- BioBriefing
- Drug Development Briefing
- Diagnostics Briefing
- BioFacilities
- Biotech 101, 201, 301
- BioPrinciples Online
- Biosafety Basics
- Biosafety Cabinets
- BioRisk Management
- Disinfection for Tissue Culture Laboratories
- NIH Recombinant DNA Guidelines
- OSHA Bloodborne Pathogens
- Working with Animals

Go to www.BiotechPrimerInc.com for a complete listing of courses, dates, and locations.

Publisher's Acknowledgements

We are proud of this book and welcome your feedback at info@BiotechPrimerInc.com.

For more information on the basic science that drives the biotech industry read our blog at www.BiotechPrimerBlog.com and follow us on twitter @BiotechPrimer. For a list of upcoming courses visit us at www.BiotechPrimerInc.com

A special thank you goes out to Emily Burke, Ph.D. who wrote and edited much of this book.

Illustrations and Cover: Michelle Leveille, ArtiFact, www.ArtifactGraphics.com

Table of Contents

TABLE OF CONTENTS

What The Heck Is Biotech?

For thousands of years—dating back to 3000–4000 B.C. when ancient Egyptians and Sumerians used fermentation to make bread, cheese, and beer—humans have employed the underlying principles of biotechnology to improve quality of life through innovations in food and medicine. In 1919, the Hungarian agricultural engineer Karl Erekya envisioned an age of biology-based revolution akin to the Stone and Iron Ages and coined the term "biotechnology". However, it wasn't until 1953 when Watson and Crick solved the structure of DNA that the modern science of biotechnology emerged. Two decades later, researchers figured out how to transfer DNA between organisms, and it was this new technology of **recombinant DNA** that marked the birth of modern biotechnology.

> **Biotechnology is the use of cellular and biomolecular processes to solve problems and make useful products.**

The impact of modern biotech has been substantial and far-reaching. In agriculture, biotech crops now have greater yields and nutritional value. In healthcare, blockbuster antibody and enzyme therapies for cancer and autoimmune disorders are saving and prolonging lives. In industrial manufacturing, new classes of biocatalysts are reducing pollution and converting plant materials into renewable biofuels. What computers were to the 20th century, biotech will be to the 21st century.

Today we have a working definition of biotechnology that, like the science itself, is a product of our advances over the past hundred years. **Biotechnology** *is the use of cellular and biomolecular processes to solve problems and make useful products.* While ancient peoples were able to make use of yeast cells for fermentation, it was not with an understanding of the underlying principles of biology. In contrast, today's scientists have a better understanding of the cell and its metabolism, allowing them to intentionally manipulate biological processes to create products that improve human life.

GOING FURTHER

Let's take a closer look at our definition of biotechnology. First, biotechnology uses many types of cells (bacterial, mammalian, yeast, insect, and plant) to manufacture products. Second, biotechnology products are themselves made up of cells or cell parts such as proteins, DNA, and RNA— each of which will be discussed in more detail in later sections. A specific example of a biotechnology product is the protein insulin, normally produced by cells in the pancreas. People with Type I diabetes do not produce insulin, so they must inject human insulin to control their blood sugar. Companies such as Genentech have developed a process to manufacture human insulin in bacteria. This is modern biotechnology at work.

Over the past several decades, the growth of biotechnology applications has resulted in the emergence of distinct industry sectors. In the following pages, we will give an overview of these major sectors: Healthcare, Agriculture, and Industrial/Environmental.

Healthcare Sector

Products in biotech's healthcare sector can be subdivided using the United States Food and Drug Administration's (FDA) categories: **pharmaceuticals**, **biologics**, **medical devices**, **diagnostics**, and **combinations**.

Pharmaceuticals are chemical compounds called **small molecule** drugs because relative to their biotech counterparts, they are extremely small in size. So small, in fact, that most can enter cells in order to fight disease. Pharmaceuticals are man-made. Some pharmaceuticals are first discovered in natural products, extracted, and then chemically synthesized for mass production. A classic example of this is aspirin, as depicted in a series of illustrations on the next page. For centuries, people have appreciated the pain-relieving properties of the willow tree bark. In the mid 1800's, chemists isolated the compound that provided this relief, salicyclic acid. In 1887, chemists at Bayer began investigating acetylsalicylic acid, a less irritating chemical derivative, and the company began marketing it as aspirin in 1899.

Biologics is the general name given to a wide group of products including therapeutic proteins, vaccines, organ and tissue transplants, and stem cell therapy. A key distinction between biologics and pharmaceuticals is in how they are made: biologics are made in cells or living organisms as opposed to being synthesized in the lab. The majority of biologics discussed in this book are protein therapeutics produced by living cells. The complexity of the particular protein being made will dictate what type of cell can be used for its production. For small, simple proteins, such as insulin, bacterial cells are used. For larger, more complex proteins, such as monoclonal antibodies, mammalian cells are used.

Biologics, an FDA term, are often referred to as **large molecules**. When compared to pharmaceuticals, biologics are enormous in size—contrast the size of a mouse (pharmaceutical) to that of an elephant (biologic). Because of their size, biologics cannot enter cells, meaning that they have to cause their therapeutic effect from outside of the actual cell. We will discuss how this is possible in later chapters.

Some pharmaceuticals are first discovered in natural products

CHAPTER ONE: WHAT THE HECK IS BIOTECH?

The development of biologics as human therapeutics was pioneered by small private companies—collectively called Biotech—that came into being in the mid 1970's. In addition to developing a new type of drug—biologics—these companies developed a new approach to drug development, referred to as **rational drug discovery** or **mechanism-based drug discovery**. Rather than randomly screening large collections of chemical compounds to see which might be effective at treating disease symptoms, researchers at these companies first sought to understand the underlying mechanisms of a given disease, and then specifically design a drug that would interfere with these processes. These **drug candidates** would then advance to testing in animal models and potentially human trials.

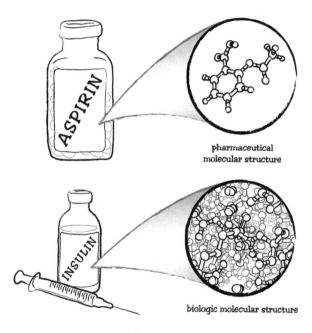

pharmaceutical
molecular structure

biologic molecular structure

Today, the line between "Big Pharma" and "Small Biotech" is blurred, as biotech companies often partner with larger pharmaceutical companies in order to manufacture and market their products, and traditional pharmaceutical companies often utilize a mechanism-based approach to drug design. Blurring the distinction still further is the fact that most traditional pharmaceutical companies now also produce biologic drugs, and some biotechs include small molecules in their product portfolios. These trends have given rise to the term **biopharmaceuticals**, used to refer to any drug made using biotech methods. The terms **life sciences** or **biosciences** encompasses both traditional pharmaceutical companies and biotechs.

Medical devices are products used for medical purposes in diagnosis, therapy, or surgery. In contrast to biopharmaceuticals, which exert a biochemical effect on the patient's body, medical devices tend to exert a physical effect. Devices include a wide range of products that vary in application and complexity—from tongue depressors to cardiac pacemakers. Rapid advances in electronics and engineering technologies continue to bring exciting new innovations to the medical device arena.

Diagnostics are tests that identify a condition, disease, or predisposition to a disease. They are performed in doctor's offices, hospitals, reference laboratories, and many are now available in kit form for home use. Diagnostics can access a patient's anatomy (mammograms looking for tumors in breast tissue), protein levels (antibody testing for HIV), and DNA content (genetic testing for cystic fibrosis in newborns). The diagnostics market is exceedingly large, and is expected to increase even further as more information regarding genetic associations with various diseases is obtained.

Combinations are a relatively new FDA category. It is the combination of any of the previous mentioned categories. An example of a combination product that consists of a device with a pharmaceutical is a drug releasing stent. These are drug coated, scaffolding devices placed in coronary arteries to help keep them open. The drug coating prevents scar tissue from forming. The home pregnancy test is an example of a diagnostic combined with a biologic: an antibody (protein) detects the pregnancy hormone human chorionic gonadotropin (hCG) to identify the pregnancy status of a woman.

GOING FURTHER

There are two major forms of diabetes—Type I diabetes, characterized by decreased or nonexistent insulin production due to the autoimmune destruction of the insulin producing pancreatic beta cells and Type II diabetes, characterized by body tissue that is insensitive to the insulin produced.

Why is it important for a diabetic to monitor their blood sugar levels closely? High blood sugar levels lead to dehydration, as the body attempts to flush out the excess sugar in urine; and life threatening changes in blood pH, as the body turns to fatty acids for delivery of energy. Diabetes has severe long term effects, including damage to blood vessels and nerves, infections of the skin and urinary tract, kidney disease, and impotence. It is one of the major chronic diseases in the industrialized world.

Cocktail Fodder

100% of Type I diabetics require insulin injections. 20% of Type II diabetics can manage their disease by changes in lifestyle alone. Type II is the more common form of diabetes.

PRODUCTS AT WORK: Type I Diabetes Management

Let's examine the development of insulin as a biologic drug as a specific example of how researchers who understand cellular function in terms of metabolism can create a life saving medication.

After eating a meal, the beta cells of the pancreas detect an increase in blood sugar, causing them to release the protein hormone insulin. Insulin is secreted into the blood stream where it interacts with receptors on other cells, especially liver, muscle, and adipose (fat) cells. The insulin signals these cells to take up any excess sugar. In liver and muscle cells, this sugar is stored as a complex sugar called glycogen. In adipose cells, it is converted to fat. This creates energy reserves that can be broken down into glucose for release into the bloodstream as needed.

Diabetes is a disease caused by the improper functioning of this system, and is characterized by high blood sugar levels. You might think having high sugar levels would be advantageous because it provides more energy for the body, but if the level of glucose remains elevated for an extended period of time then the glucose becomes toxic to the body, resulting in vascular, nerve, and other complications. Insulin production is an example of how a specific cell type, the beta cells of the pancreas, function by responding to stimuli in the environment.

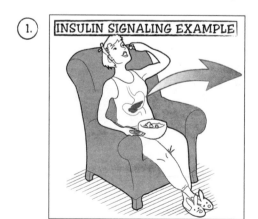

(1.) **INSULIN SIGNALING EXAMPLE**

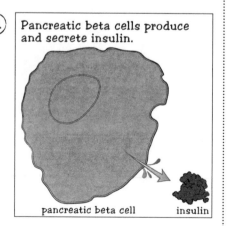

(2.) Pancreatic beta cells produce and secrete insulin.

pancreatic beta cell   insulin

Products At Work Continued From Previous Page

 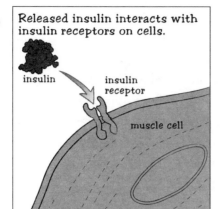

3. Released insulin interacts with insulin receptors on cells.

insulin

insulin receptor

muscle cell

 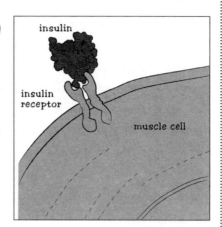

4.

insulin

insulin receptor

muscle cell

 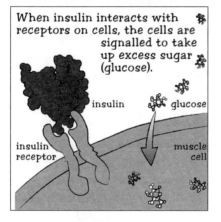

5. When insulin interacts with receptors on cells, the cells are signalled to take up excess sugar (glucose).

insulin

glucose

insulin receptor

muscle cell

So how has biotechnology improved life for diabetics? The cause of diabetes has been known for over 100 years and beginning in the 1920's, patients were given insulin purified from the pancreas of pigs and cows. Although this therapy worked for some patients, there were problems. Many people had adverse reactions to this type of insulin due to slight differences in protein structure between the human form of the protein and the cow or pig version. And as the diabetic population grew, so did concerns over having an

Products At Work Continued From Previous Page

adequate supply of purified insulin. In the 1970's, the techniques of genetic engineering allowed scientists to place the human insulin gene into bacteria so the bacteria could manufacture the human insulin protein. Today, large amounts of human insulin are efficiently produced in this way.

Next came the development of portable insulin pumps in the 1980's, and more recently insulin pens, for convenient and effective delivery of insulin. The 1990's brought major improvements in the field of medical device technology, allowing for dramatic reductions in the size of insulin pumps which made them easier for patients to use.

In recent years, insulin pumps have advanced further still by being incorporated into artificial pancreas systems. These systems consist of an electronic device capable of continuous blood glucose monitoring coupled with the insulin pump itself, which will release insulin or suspend insulin release based on feedback from the glucose monitor. A number of such systems are either in development or on the market.

Of course, the ultimate treatment for Type I diabetics will be to successfully replace damaged pancreatic beta cells with healthy functioning pancreatic beta cells. This goal may not be too far off—there have been some encouraging results from clinical trials using human stem cell therapy to replace damaged pancreatic cells.

Agricultural Sector

Farmers have been using genetic modification to produce desirable characteristics in plants and animals for thousands of years. In fact, all domesticated plants and animals have been genetically modified through the tradition of crossbreeding. The hope is that mating of exceptional plants (or extraordinary animals) will produce progeny displaying the enhanced characteristics of their parents. While crossbreeding involves the random transfer of characteristics within the same species, the techniques of modern biotechnology enable farmers to specifically transfer one or more genes from one organism to another; in some cases, between organisms of different species.

By recent United Nations estimates about 850 million people—close to one in six people in the world—do not have access to an adequate food supply. As the world's population grows, particularly in developing nations, the acreage of farmable land becomes more and more limiting for food production. The major goals of agricultural biotech are to increase yield and nutritive value of foods to adequately feed the world. Agriculturists approach the challenge of boosting crop production by genetically enhancing plants to be more resistant to weeds, insects, infection, and environmental conditions. In addition, scientists are successfully improving crop yield directly, meaning more fruits or vegetables per unit area.

Tricky Terminology

In agricultural biotechnology, both crops and livestock are referred to as **genetically modified organisms (GMOs or GM food)** or **genetically enhanced organisms (GEOs or GE food)** or genetically engineered organisms or transgenic organisms—all terms are used interchangeably.

PRODUCTS AT WORK: Agricultural Biotech

Scientists improve the nutritive value of foods through recombinant DNA technology. Over a hundred million children in developing nations have vitamin A deficiency, and one to two million will die as a result. The latest version of a plant called Golden Rice incorporates two genes that considerably increase the production of beta carotene, the precursor to vitamin A. Ongoing development efforts to improve Golden Rice include increasing the levels of vitamin E, iron, and zinc, and to improve protein quality.

What if we take nutrition a step further and consider incorporating medicines in foods? The proteins lactoferrin (not yet commercialized) and lysozyme, both used to combat infant diarrhea—a leading cause of infant mortality in the developing world—can be produced in genetically engineered rice. Growing therapeutic proteins or vaccines in foods will be a great advance, but presents interesting challenges. In creating fruits engineered to contain vaccines, for example, some considerations researchers will have to address include controlling dosage, making fruits consistent in size, and making vaccine containing fruits easily distinguishable from regular fruits.

Industrial and Environmental Sector

The industrial and environmental sectors are usually paired together because industrial processes, which often create pollution, directly affect the environment. The primary goal of both sectors is to reduce pollution. Industrial biotech achieves this by replacing energy and chemical intensive manufacturing processes with sustainable enzyme-mediated processes. **Enzymes**, which act as catalysts to speed up or slow down chemical reactions, are proteins naturally found in cells. Many of the economically valuable enzymes are mined in the world's most extreme environments, including hot springs, deep-sea vents, and the arctic poles, from microbes called **extremophiles**. Characteristics of these unique organisms include resistance to extremes of temperature, pressure, salinity, acidity, and toxins. These characteristics are encoded in their genes, which can be transferred to more conventional microbes using recombinant DNA technology. The use of enzymes can increase manufacturing productivity, reduce energy consumption, and reduce waste production, which translates into less pollution. One example of this is the heat and acid resistant catalase enzyme isolated from a thermophile, *Thermus brockianus,* which lives in deep sea hot water systems. A catalase is an enzyme that breaks down hydrogen peroxide, a common metabolic waste product in bacteria as well as an industrial pollutant. Because this catalase enzyme is heat and acid resistant, it can be used to treat hydrogen peroxide waste generated in industrial paper and fabric bleaching. Other industrially useful products from extremophiles include enzymes that enhance detergents, enzymes that digest oil, and enzymes that can convert plant matter into bioplastics.

> ## Cocktail Fodder
>
> Identification and removal of land mines is a global problem. Scientists have been able to engineer watercress plants to change color when their leaves encounter a biological agent in the air or when their roots encounter chemicals, such as nitrogen dioxide, released from land mines in the soil.

Company Close-Up

In the early 1970's, the development of recombinant DNA technology led to the founding of Genentech, the first modern biotech company. Since then Genentech has evolved to become one of the most successful biotech companies, serving as a model for many fledgling biotechs. Looking more closely at Genentech's development helps illustrate how a biotech company can achieve success in the healthcare market. Since Genentech is now in its fourth decade, we'll divide its history into four quarters, each corresponding to a particular aspect of a biotech company's growth.

The first quarter of Genentech's evolution depended on scientific innovation and discovery, and it all started with a lunch meeting between scientists Herbert Boyer and Stanley Cohen on Waikiki Beach, Honolulu, where they were at a conference on bacterial DNA. Boyer, at the University of California, San Francisco (UCSF), had found a way to precisely cut and paste pieces of DNA together; and Cohen, at Stanford University, had found out how to introduce pieces of bacterial DNA into other bacteria. Together they came up with the idea of splicing human genes into bacterial DNA and incorporating the hybrid DNA into host bacteria, enabling the host bacteria to make human proteins. In short, they invented the concept of recombinant DNA, the technology upon which the biotech industry is founded.

Once Boyer and Cohen had established the technique, Boyer was approached by an entrepreneur named Robert Swanson, and in 1976 Boyer and Swanson founded a company together based on the application of recombinant DNA technology. They called the company Genentech. In 1977, Genentech scientists produced the first human protein, somatostatin, an inhibitor of human growth hormone, in bacteria. A year later, Genentech and UCSF scientists were able to experimentally produce large quantities of human somatostatin in bacteria. Genentech scientists repeated the process yet again with the gene for human insulin. By pioneering these techniques, Genentech set a new standard for innovative discovery, research, and collaboration with academic scientists.

The second quarter of Genentech's evolution involved making and marketing biologics for human diseases through recombinant DNA technology. In 1980, as a first step in raising capital to fund drug development, Genentech held an initial public offering of its stock, raising $35 million. Shares leapt from $35 each to a high of $88 in less than an hour—the largest stock run-up ever seen in the financial markets at that time. Genentech was still too small to manufacture and market biologics on its own, so in 1982, it licensed recombinant human insulin to the pharmaceutical giant Eli Lilly. Recombinant human insulin was the first biologic drug to be sold worldwide. By 1985, Genentech had realized sufficient assets and infrastructure to produce and market Protropin, recombinant human growth hormone as a treatment for growth hormone deficiency. This was the first biologic to be both produced and marketed by a biotech company. Two years later, Activase, a recombinant form of a human protein that breaks down blood clots, was approved by the FDA as a treatment for acute myocardial infarction, or heart attack. This demonstrated the versatility of recombinant DNA technology—using biologics as a treatment, as well as a replacement therapy for missing proteins—and marked the start of Genentech's therapeutic diversification.

In its third quarter, to position itself for future development, Genentech struck a number of major deals and partnerships. In 1990, Genentech merged with the Swiss pharmaceutical company Roche in a deal worth $2.1 billion. Genentech's corporate identity and operating procedures remained in place after the merger. In 1995, Genentech entered into an agreement to extend Roche's option to purchase the outstanding company stock at a predetermined price. As part of the agreement, Genentech began receiving royalties on European and Canadian sales of some Genentech products, and Roche took charge of overseas marketing efforts.

In the mid 1990's, Genentech formed a partnership with another biotech company, Biogen Idec (then IDEC), to further develop and market a recombinant monoclonal antibody, Rituxan, that IDEC had initially developed. The partnership was rewarded in 1997 with the FDA approval of Rituxan indicated for the treatment of non-Hodgkin's lymphoma. This was the first recombinant monoclonal antibody therapeutic to be approved for the treatment of cancer. This signified Genentech's move from start-up to Big Biotech.

In 1999, Roche reoffered 22 million Genentech shares, which was the largest public offering in healthcare history at the time. When the bell rang at the end of the day's trading on Wall Street, the stock closed at $127 per share, up from $97 per share. After this, Genentech traded under the appropriate symbol DNA, matching their address—1 DNA Way. Two more offerings were made by Roche over the next year, the last raising the stock to $163 per share.

Cocktail Fodder

Rituxin was first approved in 1997 for the treatment of non-Hodgkin's lymphoma. The drug was subsequently granted marketing approval for the treatment of rheumatoid arthritis (2008) and chronic lymphocytic leukemia (2010).

Now in its fourth quarter, Genentech has well established its position in the healthcare industry. It manufactures and markets a number of biologics including Rituxan, Herceptin, Xolair, Avastin, and Raptiva. Many of these were first-in-class recombinant therapeutics, solidifying Genentech as an innovative leader in three areas—immunolongy, oncology, tissue growth and repair.

Genentech owns and operates a number of biologics manufacturing facilities in the U.S. In the mid 2000's, they sold their manufacturing plant in Spain to Lonza but entered an agreement in which Lonza would manufacture Avastin for Genentech at this site and other products at a site in Singapore. Over the past several years Rituxan, Herceptin, Activase, and Avastin received FDA approval for additional indications, demonstrating the versatility and longevity of biologic therapeutics. Also in the 2000's, Genentech received approval for Tarceva, a small molecule drug, for cancer treatment. This addition to Genentech's therapeutic portfolio reinforced its strength in oncology and demonstrates its drug development flexibility for both large and small molecules. In 2009, Roche purchased Genentech for $45 billion. Few biotechs have enjoyed the same level of success as Genentech, but Genentech's development serves as a blueprint for other biotech companies to follow.

GOING FURTHER

A biologics manufacturing facility is very different as compared to chemical compound manufacturing facilities. Why? Recall biologics are made in cells, such as bacterial or mammalian cells, which are cultured in large containers called **bioreactors**. The capacity of a bioreactor may range anywhere between 2,000 and 100,000 liters. When growing living cells on such a large scale, the task of managing such parameters as oxygen concentration and circulation, temperature, pH, and nutrient concentration of the tank becomes a complex engineering challenge.

Related Companies

Research support companies use the newest technology and innovative ideas to design, develop, and supply research tools to industry. Examples of tools provided by these types of companies include the latest in DNA sequencing machines, tools for the analysis of protein samples, as well as a whole range of products critical to **assay** development—lab-based testing of drug candidates.

Contract research organizations (**CROs**) perform research and drug development projects on a contract basis. CROs carry out many types of projects, but often specialize in preclinical and clinical studies. An emerging type of CRO, specialized CROs, provide an important service in clinical trials—ensuring that new drugs are tested on genetically diverse populations. For example, a specialized CRO might coordinate minority focused clinical trials. More than a third of the U.S. population is represented by ethnic minorities, yet minorities have been traditionally underrepresented in clinical trials.

Academic, Government, and Non-Profit Research

Although private companies spend a significant amount of money on research, the largest part of the knowledge base comes from government and non-profit research institutions. Why? Company research efforts are focused on immediate and realizable goals—making a new drug or improving a marketed drug, for instance—while publicly funded research is concerned more with answering fundamental questions in biology rather than achieving a marketable endpoint. In other words, the driving force behind innovation and discovery in biological science comes very much from bottom-up, basic research. The commercial sectors, of course, benefit from this discovery research and help to put into clinical practice many of the discoveries.

Researchers in these non-profit sectors publish their results in peer reviewed papers, such as general science, high profile journals like *Nature* and *Science* or more specialized journals such as *Journal of Virology*. Data are also deposited in public databases, such as Genbank, which is hosted by the National Center for Biotechnology Information (NCBI) at The National Institutes of Health (NIH). Since these publications and databases are publicly available, researchers at biotech companies are able to access this information and use it as a starting point for new projects.

Because biotech and pharmaceuticals company researchers can directly benefit from the basic science research conducted in academic labs, they often fund research at universities as part of collaborative agreements. The company does not usually "own" the research results obtained at the university, but the agreement will generally stipulate some kind of intellectual property rights or right of first refusal to the company. An example of this is the 2007 Biogen Idec collaboration with the Brain Science Institute (BSI) at Johns Hopkins University (JHU). The BSI is a newly created interdisciplinary institute that brings together leading researchers in neurology research from across JHU's departments. The collaboration involves research into neurodegenerative diseases like multiple sclerosis, Alzheimer's, and Parkinson's disease. It is being led by the institute's director, John Griffin, who is also a member of the Biogen Idec scientific advisory board. Biogen Idec will acquire intellectual property access in exchange for supplying research funding to the Institute.

EXCERPT FROM BIOTECH PRIMER BLOG

Biofuel From Proteins?

Currently, biofuel production relies on the conversion of carbohydrates (sugars) or fats found in plant based material such as corn or algae into energy such as ethanol. Today, most ethanol is derived from corn stock, but serious efforts are underway to make algae the primary source of biofuels, due to the better economies of growing algae.

Plant sugars are found in the cellulosic parts of the plant—making the sugars themselves difficult to access, as the cellulose must first be enzymatically digested or otherwise treated. In addition, carbohydrates make up only about 10% of the mass of a typical algae strain, with fats making up at best 25% of the mass. This means that in a typical algae, proteins actually make up 65% or more of the biomass—biomass that is not available to be converted to energy.

This may soon change. Researchers at UCLA have genetically engineered a strain of algae to produce protein that can be utilized for energy production. The same research team, led by James Liao, has also published work describing the creation of an algal strain capable of directly converting cellulose to isobutanol—eliminating the need to first extract sugars from the cellulose. Although both of these findings will need to undergo extensive developmental work before they are ready for commercial use, they herald an exciting future for alternative biofuels.

http://www.BiotechPrimerBlog.com

In this chapter, we've defined biotechnology, outlined its major sectors, and described the development of a leading biotech company. In the next few chapters, we'll explore the basic science and technology that make the industry possible.

CHAPTER TWO

Putting The BIO In Biotech

At its most basic level, biology seeks to understand how cells function—what they do, and how they do it. Biotechnology then applies this understanding towards the development of new products—from cancer treatments to biofuels. In later chapters, we will examine in more detail some of these literally life changing products. In this chapter, we will build the foundation that makes these products possible.

Cells: The Basic Unit of Life

All living things are made of cells. Organisms can be single celled, also known as **unicellular**, such as yeast or bacteria, or they can be **multicellular**, such as plants, animals, and people. Most people are much more familiar with multicellular organisms—their children, their pets, and plants—but as we'll see, unicellular organisms are of great importance to the biotechnology industry.

Cocktail Fodder

The ostrich egg holds the distinction of being the world's largest single cell.

There are two distinct classes of cells: prokaryotic cells (**prokaryotes**) and eukaryotic cells (**eukaryotes**). You may never have heard of prokaryotes before—but they are everywhere! All bacteria fall into this category, and just about everything you see around you is covered in bacteria. No need to be alarmed—most bacteria do not cause disease, and in fact may even be beneficial, such as the bacteria that reside in your gut to help you to digest food.

But where does that unusual name—prokaryote—come from? Breaking it down, we can note the prefix "pro", which means before. "Karyote" derives from the word "karyon", an anglicization of the Greek word for "nut" or "kernel"—signifying the cell nucleus, a specialized structure in the middle of more complex cells that houses the genetic material, or DNA. So prokaryotic simply means "cells without a nucleus". Nearly all non-bacterial cells fall into the other broad category of cells,

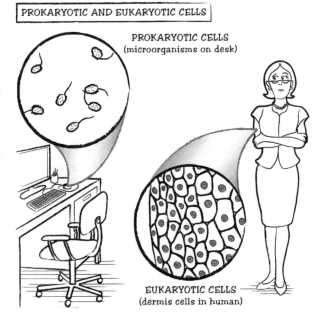

PROKARYOTIC AND EUKARYOTIC CELLS

PROKARYOTIC CELLS
(microorganisms on desk)

EUKARYOTIC CELLS
(dermis cells in human)

the eukaryotes. The prefix "eu" simply means true—so all eukaryotic cells contain a "true nucleus". Every cell that makes up your body—and the bodies of all other multicellular organisms—is a eukaryotic cell.

GOING FURTHER

The adult human body is made up of roughly 100 trillion cells—that's 100 million, million—of more than 200 different types, including organ specific cells, muscle cells, and nerve cells that were all derived from one single, fertilized egg. Each of these many different types of cells is designed to carry out a specialized function, yet they work together in concert 24 hours a day, 7 days a week, 365 days a year for up to 80-90 years. Now that's impressive!

INDUSTRY NOTE

A common bacteria found in your intestines, *Escherichia coli* (*E. coli*), is the workhorse of bacteria in research labs. Because *E. coli* has been so thoroughly studied and is approved by the FDA for the production of human therapeutics, it's also the most widely used prokaryote in the biotech industry. You may have heard that *E. coli* is a disease causing bacterium. Only certain strains of the *E. coli* bacteria cause disease; the rest are harmless.

Cell Structures

We now know all living things are made up of cells and there are two major classifications of cells. Let's look more closely at the structures found within these cells and what the functions are for each structure.

All cells are surrounded by a **cell membrane**, which keeps the cell's contents inside and the cell's surroundings outside. The membrane acts with discretion to allow some molecules into the cell; for instance, sugars enter to provide energy. Equally important, the membrane keeps out other molecules that would be harmful to the cell. It also allows metabolic waste products to exit the cell. It is very flexible, allowing the cell to change shape and move freely. Some cells, such as bacteria and plant cells, have an additional barrier called a **cell wall**, which adds further structural protection.

In prokaryotes, the genetic material of the cell, its DNA, floats around inside the cell, but in eukaryotes—yeast, invertebrates, mammals—the DNA is more safely housed in the nucleus. DNA provides a blueprint for making the proteins that perform all

Cocktail Fodder

Ever wonder why you don't melt when you shower? Didn't think so! If you had, the answer is your cells' membranes are made mostly of fat, which is hydrophobic (water resistant).

the cell's functions. The nucleus is surrounded by its own membrane, the **nuclear membrane**, which provides an additional layer of protection for the enclosed DNA.

Almost all cellular functions are carried out by proteins. And as we shall see, many biotech products are also proteins. Proteins are made by **ribosomes**—the cell's protein factories.

Making proteins, allowing molecules to cross the membrane, and all of the other functions carried out in the cell requires energy. Energy is supplied by **mitochondria**, the cell's power plants, which take the energy contained in the chemical bonds of sugars and fats and convert it into a form of energy the cell can use—a molecule called **ATP**.

Cocktail Fodder

Mitochondria have their own DNA. Humans receive 100% of their mitochondrial DNA from their mothers.

The cell manufactures thousands of different types of proteins, and many of them have to get to a particular place in the cell to do their job—the membrane, for instance. The **Golgi body** acts as a post office to the cell—sorting each protein and sending it to its correct destination.

Organelles and proteins often need to move within the cell, and the cell itself needs to move or rearrange itself, especially when it divides. Transport within the cell and movement of the cell is coordinated and carried out by an extensive network of tubes and filaments known collectively as the **cytoskeleton**, which acts as the cell's freeway. The cytoskeleton also serves as scaffolding for the cell membrane—literally, a cellular skeleton.

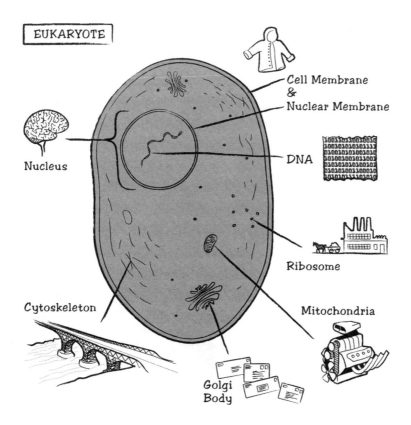

EUKARYOTE

Cell Membrane & Nuclear Membrane

Nucleus

DNA

Ribosome

Cytoskeleton

Mitochondria

Golgi Body

GOING FURTHER

It's not surprising the heart and other muscles contain a large number of mitochondria in their cells since they have a great energy requirement. Another organ that uses substantial amounts of energy is the brain, and a whopping 20% of your glucose intake ends up there! Researchers have found the brain cells of people with neurodegenerative diseases such as Parkinson's disease and Alzheimer's disease do not contain as many mitochondria as those of healthy people. This presents the possibility of a new therapeutic approach—developing drugs that are able to preserve mitochondria in brain cells.

But What Do Cells *DO?*

Of course, the precise answer to that question depends on the particular cell type we're talking about. Heart cells pump blood, liver cells detoxify chemicals, pancreatic cells make insulin, etc. However, all cell types, prokaryotic and eukaryotic alike, no matter what their specialized function is, carry out three basic functions: communication, growth and division, and manufacturing of proteins. Let's take a closer look at each of these functions.

Cells Communicate

In multicellular organisms, cells must communicate with each other. Since cells don't have mouths, ears, or access to email, they must rely on chemical messengers. A chemical message—for example, a hormone—is released by one cell, and received by a second cell—the target cell. The target cell receives the message through proteins inserted into its membrane known as **receptors**—proteins that control the passage of molecules and the flow of information

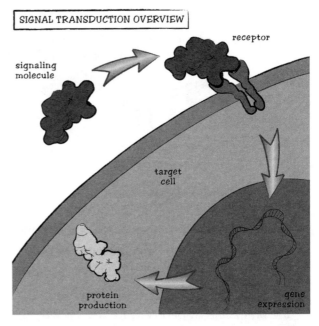

across the membrane. When the signaling protein binds its receptor, the receptor changes shape and **transduces**—converts from one form to another—the chemical message across the membrane to the cell interior. This process of cellular communication is known as **signal transduction**. The most common end result of signal transduction, and a key step in cell decision making, is the switching on, or off, of protein production—more commonly called **gene expression**. We will more fully explore gene expression in later chapters.

Another class of membrane proteins that aid in cellular communication is **channel proteins**. These proteins act as molecular gates that allow the passage of small molecules and ions, for example, glucose and sodium, across the membrane in response to a stimulus, such as insulin signaling in the case of glucose or an electrical current in the case of ions. In neurons, ion transport between cells serves as a principle means of signal transduction. The influx of calcium ions (Ca⁺⁺) into a neuron results in the release of neurotransmitters—chemical messengers specific to the nervous system. Different types of neurotransmitters regulate a variety of brain functions, including muscular activity, memory, learning, and mood regulation. Communication between nerve cells will be examined in more detail at the end of this chapter.

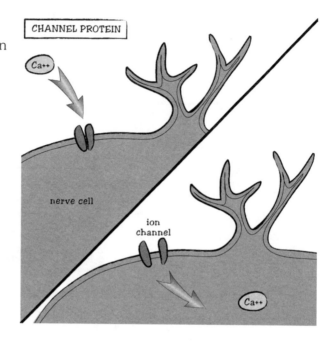

GOING FURTHER

As described in Chapter One, the regulation of blood sugar levels by the protein hormone insulin is an example of cellular communication. After you eat, beta cells in your pancreas sense increased blood glucose and respond by releasing insulin into the bloodstream. Insulin molecules attach to specific insulin receptors on muscle cells and, in doing so, deliver a signal to the inside of the muscle cell to send glucose channels to the membrane, resulting in glucose uptake. In this way blood glucose levels are kept constant, which means that the cell maintains **homeostasis**.

Cells Grow and Divide

All cells can grow in size. As a cell grows, it makes more organelles, such as mitochondria, to keep up with energy demand. The cell also produces proteins as it grows—in fact, proteins are the main source of the increased mass.

Before a cell divides, it must copy its entire DNA sequence—its **genome**—so one copy goes into each new cell, called a daughter cell. When a cell divides, each daughter cell possesses one copy of the original genome, as well as approximately the same number

NORMAL CELLS vs. CANCER CELLS

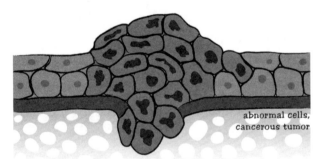

normal cell tissue layer

abnormal cells, cancerous tumor

of organelles, and the daughter cell eventually grows to the same size as the original parent cell.

The processes of cell growth and division are tightly linked and strictly controlled. Some cells, like those in the skin or lining the gut, grow and divide often. Most cells in a mature organism, however, grow and divide rarely, if at all. Cells will grow and divide ONLY if they receive a growth signal from outside of the cell called a growth factor. When growth factors are released and attach to the growth factor receptor in the cell membrane, a cascade of signals "activates" the cell's DNA, resulting in the production of proteins required for cell division. A hallmark of cancer cells is that they lose the requirement for a growth signal and continue to grow and divide in an uncontrolled manner. See *Cell Signaling: A Closer Look* on page 31 for a more detailed description of the signaling pathways involved in cell growth and division.

GOING FURTHER

Damaged or unnecessary cells are removed from the body by the process of programmed cell death known as **apoptosis**. Understanding the mechanisms underlying apoptosis may lead to discovering new treatments for cancer—a disease caused by cells continuing to grow and divide even when they should not.

Cells Manufacture Proteins

From the perspective of the biotechnology industry, one of the key properties of cells is their ability to manufacture biologically important molecules, especially proteins. This is because many biotech products, as we shall see, are actually proteins made by cells.

Cocktail Fodder

Ribosomes are able to assemble an average sized protein in about one minute.

The instructions for how to make the proteins are supplied to the ribosomes by the cell's DNA. When supplied with sufficient energy and information, the ribosomes can accomplish their manufacturing task and produce proteins. Once they are made, many proteins physically interact with other proteins to form functional multi-protein complexes. Almost all critical cellular functions—from enabling chemical reactions to communicating with the outside world—are carried out by proteins.

Cell Signaling: A Closer Look

Some cells send signals while others receive signals, but most cells do both. The signals are chemical hormones, such as adrenaline, or proteins, such as insulin. They are produced within specialized cells (the signaling cell) and released to find their target cells. The signal is often called a **ligand**. In some cases, the signaling cell and target cell may be the same cell.

The target cell may be in direct contact with the signaling cell, or it may be in a different part of the body and receive a signal that has been transported through the bloodstream. Alternatively, the signal and target may be in close proximity and the signal can be transported by diffusion through the intracellular space. After receiving a signal, the target cell responds in a manner that is determined by the nature of the signal received.

In the next few paragraphs, we will take a look at three key types of cell signaling: growth factor signaling, G protein coupled receptor signaling, and ligand gated ion channel signaling.

(1.)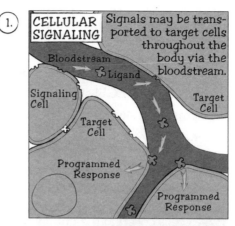

CELLULAR SIGNALING — Signals may be transported to target cells throughout the body via the bloodstream.

(2.)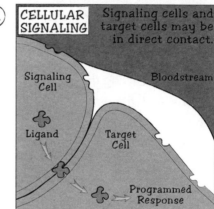

CELLULAR SIGNALING — Signaling cells and target cells may be in direct contact.

(3.)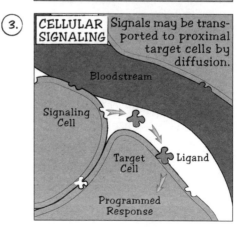

CELLULAR SIGNALING — Signals may be transported to proximal target cells by diffusion.

Growth Factor Signaling

Growth factors are proteins that signal a cell to multiply. For instance, **epidermal growth factor** (**EGF**) stimulates the proliferation of skin cells during wound repair. Cells are constantly exposed to many different growth factors, and the particular ones they respond to depends on their cell surface receptors. Skin cells, as well cells covering the gut, lung, and breast, have or express receptors for EGF, while nerve cells express receptors for **nerve growth factor** (**NGF**).

After receiving the initial growth factor signal, the enzymatic activity of the internal portion of the growth factor receptor is activated. The particular type of activity switched on is protein **kinase** activity—or the ability to transfer a phosphate group from one molecule to another. These types of receptors are sometimes referred to as receptor tyrosine kinases (RTKs), because they selectively transfer phosphate groups to the amino acid tyrosine on the recipient protein. This transfer, in turn, causes a slight shape change in the protein which received the phosphate group, typically leading to the activation of that protein's own kinase activity. This newly activated

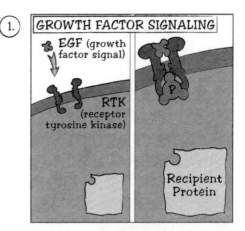

(1.) GROWTH FACTOR SIGNALING

EGF (growth factor signal)

RTK (receptor tyrosine kinase)

Recipient Protein

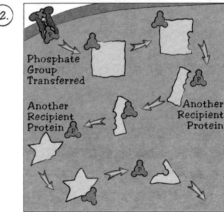

(2.) Phosphate Group Transferred

Another Recipient Protein

Another Recipient Protein

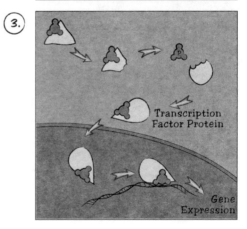

(3.) Transcription Factor Protein

Gene Expression

protein kinase then goes on to activate yet another kinase protein, and so on, in what is referred to as a signal transduction cascade. The last element in this cascade to be phosphorylated is typically a protein called a transcription factor. Once phosphorylated, the transcription factor enters the nucleus, where it binds to the DNA at a particular location, activating expression of a specific gene.

Defects in the growth factor signaling process are associated with different types of cancer. A major challenge in oncology lies in understanding the complex signaling pathways that trigger cell division and determining what has gone wrong in each type of cancer. Once these signaling pathways are understood, it is possible to develop targeted therapies for the particular cancer.

G Protein Coupled Receptor Signaling

A common type of receptor molecule, widely used in almost all cells to regulate a variety of processes, from blood pressure to nerve transmission to stomach acid secretion, are called **G protein coupled receptors** (**GPCRs**) because they are coupled to a type of signaling protein called a G protein. The GPCR is more structurally complex than a growth factor receptor, in that it crosses back and forth across the membrane seven times and is therefore sometimes called a seven transmembrane receptor or a serpentine receptor.

Upon ligand binding, the GPCR undergoes a subtle conformational change, causing it to interact with the associated G protein. This interaction causes a portion of the G protein to disassociate from the rest of the complex, and interact with other proteins within the cell. Some of these intracellular proteins are kinase proteins, as described for growth factor receptor activation.

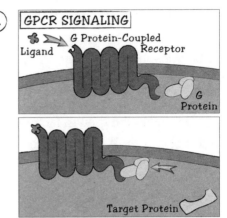

1.

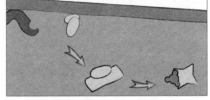

2.

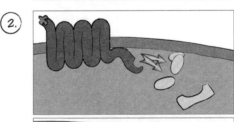

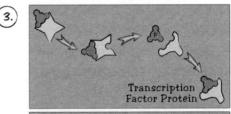

3.

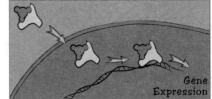

INDUSTRY NOTE

GPCRs may be activated by a wide variety of ligands, including simple chemicals such as epinephrine and histamine, or by more complex signaling proteins. The results of GPCR activation may also be quite varied, ranging from the activation of gene expression to the release of an already made protein to the reorganization of the proteins that make up the cytoskeleton. Perhaps because of this variety of responses and consequent critical role these receptors play in a variety of biological processes, they are popular drug targets. It is estimated that between 25% and 40% of all biotech drugs currently in development or on the market target GPCRs.

Ligand Activated Ion Channels

Insulin communicates signals regulating glucose homeostasis from the pancreas to different tissues all around the body. However, short range communication in the central nervous system, i.e. between nerve cells, or neurons, is critically important for a whole range of body functions and involves ion channels for signal transduction.

A ligand activated **ion channel** recognizes its ligand and undergoes a structural change that opens a gap

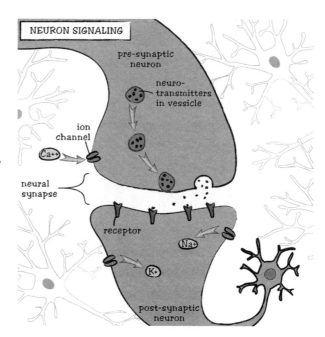

NEURON SIGNALING

pre-synaptic neuron

neuro-transmitters in vessicle

ion channel

Ca++

neural synapse

receptor

Na+

K+

post-synaptic neuron

(channel) in the plasma membrane through which **ions**—positively or negatively charged atoms—can pass. These ions relay the signal. An example for this mechanism is found in the receiving cell, or post-synaptic cell, of a neural synapse.

Nerve impulses are transmitted as tiny pockets of electrical current, called action potentials—sodium and potassium ions enter a neuron through ion channels in the neuron's plasma membrane. These ions then travel along the neuron until it gets to the end. The action potential cannot "jump" across this space, so the signal has to be translated into a different form.

When the sodium and potassium ions reach the nerve ending, they cause a set of ion channels to open in the membrane, allowing calcium ions to enter. When calcium ions flood into the nerve cell, intercellular vesicles containing signaling molecules known as neurotransmitters fuse with the plasma membrane, releasing their content into the space between nerve cells, called the synapse. These newly released neurotransmitters can then bind receptors in the next neuron, allowing sodium and potassium ions to enter, starting the process over again.

Since it is important to switch off nerve impulses once they have been transmitted, the neurotransmitter is rapidly removed from the synaptic space by specific reuptake channels in the presynaptic neuron, terminating the signal until another action potential arrives. Blocking these reuptake channels with appropriately shaped molecules therefore sustains and amplifies neuronal signaling and can have significant impact on neural function.

PRODUCTS AT WORK: Antidepressants

Lower than average levels of the neurotransmitter serotonin have been linked to depression. This observation was the foundation for a highly successful class of antidepressants, selective serotonin reuptake inhibitor (SSRI) antidepressants, including Prozac, Zoloft, and Paxil. These medications block reuptake of serotonin, thereby increasing its concentration in the synaptic space and helping to relieve the symptoms of depression. Some recreational drugs also act as reuptake inhibitors for various neurotransmitters. Cocaine, for example, blocks the reuptake of the "feel good" neurotransmitter dopamine.

Signaling Network Complexity

Although we have schematically represented signaling through linear, individual pathways, cell signaling in reality is best thought of as a network of signaling events integrating inputs from multiple sources and processing them to produce the appropriate output. So, in the growth factor RTK signaling pathway, a growth factor binds to its receptor and recruits signaling proteins to the receptor inside the cell. This initiates a chain of signaling

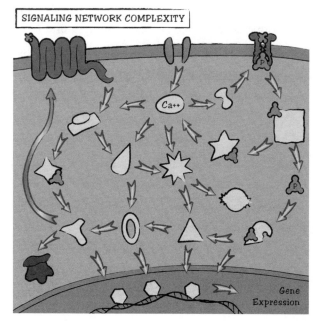

SIGNALING NETWORK COMPLEXITY

Ca++

Gene Expression

events leading to the activation of proteins that bind to DNA and switch on (or off) gene expression.

However, there are hundreds of different receptor types, encoded by different genes, on the surface of every cell, each one binding a different ligand (growth factor, hormone, neurotransmitter, etc.). They include receptor tyrosine kinases, GPCRs, and ion channels. Some receptors, like GPCR, bind more than one factor. Ligand binding to each of these receptors starts a different signaling cascade, ultimately leading to regulation of gene expression and cellular function. The important point to note is that the different signaling pathways are not isolated; there are several instances of interaction between the different pathways—for example, between the GPCR pathway and the RTK pathway, in which a signaling protein from one pathway interacts with a signaling protein from another pathway.

The result of activation of these signaling networks is the activation of a variety of transcription factors leading to an integrated gene expression response dependent on the relative strengths of each extracellular signal. Further complexity is introduced when we realize that signaling networks impact cellular events in addition to nuclear gene expression—for instance, protein synthesis at ribosomes, or the secretion of proteins that are stored within the cell. Other sites of signaling response include the cytoskeleton and mitochondria.

EXCERPT FROM BIOTECH PRIMER BLOG

Powerful Genes

Many people recall from their high school biology days that the cell's nucleus houses its DNA, or genetic material, and some will even recall that the mitochondria are commonly referred to as the cell's "powerhouse" because they carry out the chemical reactions that provide the cell with energy. Few, however, will remember that the mitochondria actually also contains its own DNA.

Scientists have so far believed that this mitochondrial DNA mostly codes for enzymes involved in carrying out the chemical reactions needed to provide cellular energy. It turns out, however, that mitochondrial DNA may also hold the recipe for proteins involved in various diseases. A paper recently published in PLoS One (DOI:10.1371/journal.pone.0006423), the open access online journal, describes a surprising correlation between specific versions of certain mitochondrial genes and a reduced risk for stroke. This finding helps to explain why previous genome-wide association studies—using only nuclear DNA—did not identify any genes associated with strokes, even though researchers had suspected a genetic link. Preliminary studies also suggest a possible link between certain mitochondrial genes and other complex diseases such as Parkinson's, Alzheimer's, and diabetes, opening up exciting new possibilities for understanding human disease.

http://www.BiotechPrimerBlog.com

In this chapter, we have described the two major categories of cells, discussed the major components of eukaryotic cells, described the three basic functions of all cells, and took a detailed look at cellular communication. In the next chapter, we will examine life's blueprint: DNA.

CHAPTER THREE
Life's Blueprint

We learned in the previous chapter that DNA is the genetic material contained within cells. Often referred to as an instruction manual or a blueprint for life, DNA is the molecule used by all living organisms to store and transfer information. In this chapter, we will look inside the cell and examine the molecule upon which the entire biotechnology industry is built, starting with some key discoveries and the scientists behind them.

Cocktail Fodder

Red blood cells in the human body are the only cells that do not contain DNA. Why? The cell can't spare the room. Each blood cell has evolved so that it can carry the maximum amount of oxygen and carbon dioxide to and from other cells. Without DNA, blood cells cannot divide, so it's up to the bone marrow to replenish blood cells approximately every three to four months.

History of DNA

In the 1850's, an Austrian monk named Gregor Mendel did breeding experiments with pea plants and observed that certain genetic characteristics were passed down from one generation to the next in specific ratios. Because he was the first person to analyze inheritance of traits in a systematic way, Mendel is considered to be the father of genetics. What makes Mendel's contributions even more impressive is the fact that even though he didn't know about DNA, he predicted the existence of what he referred to as "particles of inheritance" that were responsible for passing on traits from generation to generation.

In 1944, scientists identified DNA as these "particles of inheritance". The race was on to determine its structure, a scientific puzzle that was solved by the research team of James Watson and Francis Crick in 1953. The structure that Crick and Watson conceived was a **double helix**. DNA was shown to be helical in nature by an x-ray image provided by a scientist named Rosalind Franklin, who died before ever receiving full credit for her critical contribution. Many other possible structures had

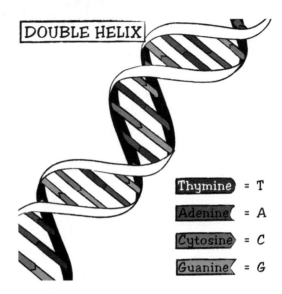

DOUBLE HELIX

Thymine = T
Adenine = A
Cytosine = C
Guanine = G

been proposed, including a triple helix, but the double helix model was the only one that fit the evidence of base pairing provided by Erwin Chargoff. In the last sentence of Watson and Crick's *Nature* paper in which they published the double helix structure, they made an important point, which we will soon revisit: "It has not escaped our notice that the specific pairing we have postulated immediately suggests a possible copying mechanism for the genetic material."

Life's Building Blocks

The full name for DNA is deoxyribonucleic acid. It's a long molecule made up of repeating subunits called **nucleotides**. Each nucleotide has three parts:

1. A sugar molecule called **deoxyribose**—the "D" part of DNA. This is a molecule with 5 carbon atoms—each of the carbons are numbered 1',2',3',4',5'—which form a ring structure.

2. A **phosphate** molecule attached to the fifth carbon atom in the ring. This is what gives DNA its weak acidic properties—the "A" in DNA.

3. A molecule called a **base** attached to the first carbon atom in the ring. There are four different bases: **adenine** (A), **cytosine** (C), **guanine** (G), and **thymine** (T).

NUCLEOTIDE

A nucleotide is a repeating subunit of DNA. Each nucleotide is composed of a phosphate group, a sugar, and a base. Here are three different ways of viewing nucleotides:

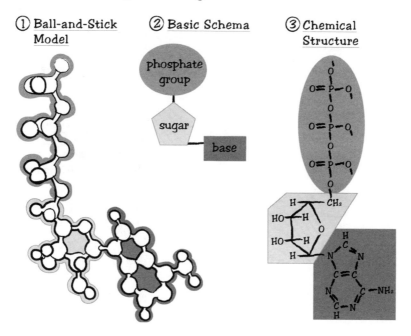

① Ball-and-Stick Model

② Basic Schema

③ Chemical Structure

The prefix "deoxy" means "without oxygen" and refers to the lack of an oxygen molecule at a particular position in the sugar molecule. "Ribo" stands for ribose, which describes the particular kind of sugar. The four bases A, C, G, and T are the only ones used to make up DNA, and all organisms use the same four bases and the exact same sugar and phosphate molecules. *In other words, the nucleotide building blocks for the genetic material of all forms of life are identical.*

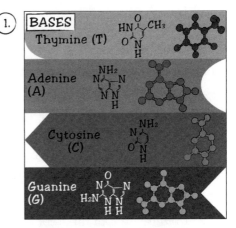

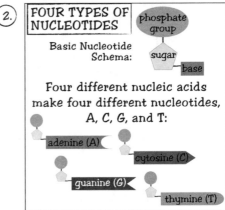

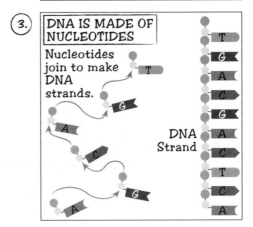

Putting the Building Blocks Together

Nucleotides are linked together to form a long chain. During DNA synthesis there is an enzyme in your cells called DNA polymerase that links the phosphate group of one nucleotide to the sugar group of another, forming what is often referred to as the sugar-phosphate backbone.

This explains how bases connect to form a DNA strand, but how do we get from nucleotide chains to a double helix structure? An early experiment by Erwin Chargoff revealed that within a DNA molecule the number of Cs was always the same as the number of Gs, and the number of Ts was always exactly the same as the number of As. This led to the conclusion that DNA is made up of two paired strands: Cs on one strand are matched to Gs on the other, and As on one strand are matched to Ts on the other. The pairing of Cs to Gs and As to Ts between strands is accomplished by chemical bonds, and the geometry of those bonds, determined by the particular shapes of the nucelotides themselves, is what gives the molecule its helical shape.

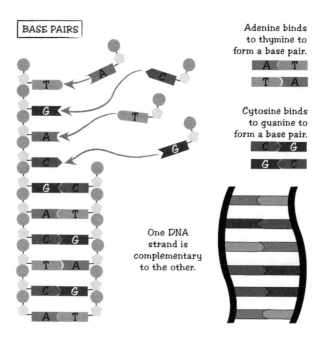

BASE PAIRS

Adenine binds to thymine to form a base pair.

Cytosine binds to guanine to form a base pair.

One DNA strand is complementary to the other.

Although all living organisms on earth have DNA made up of the same four nucleotide bases, these bases are ordered differently in each living thing. This is why all people—indeed, every living thing—are unique. They each have their own unique blueprint for life.

The sugar-phosphate bonds that form the molecule's backbone are very strong and give the information content of DNA—the order of those bases—great stability. This is why it's possible to retrieve DNA from some fossils and why police detectives are able to collect intact DNA forensic evidence from crime scenes days, weeks, or even months after the crime takes place.

Cocktail Fodder

Where do your cells get the nucleotide building blocks needed to make DNA? The food you eat, such as chicken, fish, fruits, and vegetables, are made of cells. Since animal and plant cells also contain DNA made of the same constituent parts as human DNA, the DNA you consume in foods is broken down into nucleotides and used by your cells. The four nucleotides can also be synthesized by our cells *de novo*, meaning from the beginning, using atoms from proteins, sugars, and fats that we consume.

How DNA Replicates

Recall the statement made by Watson and Crick, when they presented their model of the DNA molecule as a double helix, that the structure they proposed "immediately suggests a possible copying mechanism for the genetic material". Five years after Watson and Crick elucidated the structure of DNA, the process of DNA replication was verified. The key piece of the puzzle was the discovery of the enzyme that facilitates the process, **DNA polymerase**, by Nobel Prize winning scientist Arthur Kornberg.

Cocktail Fodder

Scientists have recovered DNA from the wooly mammoth and hope to use it to recreate the extinct animal. An elephant would act as a surrogate because of its genetic similarity to the wooly mammoth.

Although the chemical bonds that hold adjacent nucleotides together are very stable, the chemical bonds holding complementary base pairs together are relatively weak. This makes the first step in DNA replication, strand separation, easily accomplished by an enzyme called **DNA helicase**. Once the two strands are separated, each one can act as a **template** for the synthesis of a new complementary strand.

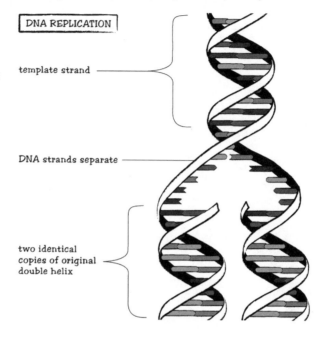

DNA REPLICATION

template strand

DNA strands separate

two identical copies of original double helix

The appropriate free floating base first pairs up to each exposed base on the separated DNA strand. DNA polymerase then catalyzes the formation of sugar-phosphate bonds between the newly aligned base pairs, creating the new complementary strand. Working one base at a time, DNA polymerase moves along each separated strand, using it as a template to synthesize its complementary strand until the chain is completed and the DNA is fully replicated. At the end of DNA replication, there are two new DNA molecules, each identical to the molecule we started out with.

GOING FURTHER

To work with DNA molecules, scientists must extract and isolate genomic DNA from living cells. There are many variations in the procedure for extracting and isolating DNA from cells, but most rely on three basic steps:

1. Lysis. Cells are broken open, or lysed, in a detergent containing solution. Since the cell membrane is made of lipids, which are fatty, "greasy" substances, it dissolves in detergent—the same way bacon grease can be washed off of a frying pan using detergent. Once the cells are broken open and their contents released, we have what is referred to as a **cell lysate**.

2. Separation from protein. The detergent that was used in the first step to break open the cell membrane serves a dual purpose—it also **denatures**, or unfolds, the proteins to which the DNA is attached, thereby helping to separate the DNA from those proteins. Most lysis buffers also contain **proteases**, or enzymes that break down proteins. These steps—denaturing and cleaving proteins—are necessary because in order for scientists to work with the DNA in subsequent steps, it must be protein-free.

3. Precipitation. Next, the DNA is precipitated in ethanol. This simply means that the addition of ethanol to the sample tube causes the DNA to form a white solid which can now be more easily isolated from the rest of the sample. Once this precipitate forms, the sample tube is centrifuged, leaving a pure DNA pellet at the bottom of the tube. This DNA pellet can then be dissolved in a small volume of buffered water, stored in the freezer, and used for later applications. In an appropriately buffered solution, frozen DNA is stable for several decades.

Organization of DNA

The genome of an organism is the full complement of its DNA. Each cell of the organism contains an identical copy of its genome. A genome is organized into chromosomes, which are packages of DNA. Chromosomes are highly condensed DNA wrapped around a protein. If we "unwrap" a chromosome, we see that it is simply a long strand of DNA. Along this strand we will find many genes. A gene is simply a segment of DNA that codes for a protein. For example, the insulin gene, found on chromosome eleven, provides the blueprint to make the insulin protein.

Different organisms have different numbers of chromosomes. Humans, for example, have 23 pairs of chromosomes for a total of 46. The exception to this is the germ line cells—the egg and the sperm cells. These cell types have 23 chromosomes—a single copy of each. When a sperm cell fertilizes an egg cell to form a zygote, the chromosome count is restored to 46.

Cocktail Fodder

Proteases are enzymes that break down proteins, and are found in abundance in some fruits such as papaya, pineapple, and kiwi—which is why some fruits can be used as a meat tenderizer.

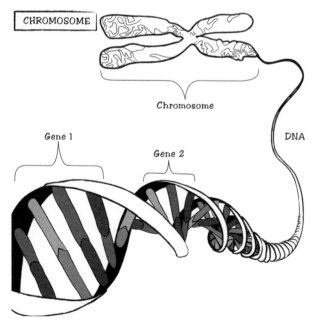

CHROMOSOME

Chromosome

Gene 1

Gene 2

DNA

The information contained in DNA—the genetic code—is read in groups of three bases, known as **codons**. Varying lengths of codon sequences make up genes and contain the information that cells use to make proteins. If we think of an organism's genome as a book, then the various components of the DNA code that we've discussed are analogous to the writing within that book. The nucleotides of the genome are its letters, the codons are its words, the genes are its sentences, and the chromosomes are its chapters. Individual words and sentences have limited meaning independently; it's the collection of chapters as a whole from start to finish that tells the story.

The Human Genome Project

In 1990, researchers at the National Institutes of Health (NIH) spearheaded an ambitious international research effort. Their goal was nothing less than determining the sequence—the order of bases—of the entire human genome. After 10 years of effort by an international consortium, a draft sequence of the human genome was published in 2000 and a more complete sequence published in 2003. In addition to the NIH-led effort, a private company, Celera Genomics, also succeeded in sequencing the human genome.

Based on data from the Human Genome Project, we know that the human genome contains between 20,000 and 25,000 genes—a much lower number than originally predicted. Even more surprising, these ~25,000 genes only make up about 1.5% of the genome. Most of the genome lies outside of the coding regions of genes. This noncoding DNA used to be referred to as junk DNA, but more recently scientists have uncovered evidence that it plays an important role in the regulation of genes and gene expression. Genomics seeks to understand the function not only of the genes, but also of the noncoding regions of the genome.

INDUSTRY NOTE

James Watson's own personal genome was published online in 2007 after it was sequenced by the biotech company 454 Life Sciences, now a subsidiary of Roche. It took two months to complete at a cost of $2 million. Watson agreed to have his genome published except for the genes associated with Alzheimer's disease because of a family history of the disease. A major goal for the biotech industry is to reduce the cost of sequencing a complete human genome to the point where it is affordable for routine clinical use. With each passing day, advances in sequencing technology bring this goal closer to being realized.

EXCERPT FROM BIOTECH PRIMER BLOG

Back to the Future

We most often think of biotechnology as the key to the future—the 21st century tool that will be used to heal, fuel, and feed humanity. However, an article published in *Science* in May 2010 (Green et al., 7 May 2010, pp. 710–722) suggests that advances in biotechnolgy also enable us to better understand our past.

In this article, Richard Green and colleagues at the Max Planck Institute in Liepzig, Germany, published the complete DNA sequence—the genome—for Neanderthals. They were able to do this using DNA recovered from fossils found in a Croatian cave. Using this information, they were then able to determine at least 15 different genes that clearly differentiate modern humans from Neanderthals. This included genes involved in mental development, metabolism, and in developing parts of the human skeleton.

Although this discovery may not have an immediate impact on curing disease or feeding our growing population, it does offer a unique glimpse at our evolutionary past. And this glimpse was made possible by the most modern of technologies for extracting and deciphering—sequencing—DNA. The Neanderthal bones were not particularly well-preserved, so new techniques were required to extract usable DNA from the bones, and to differentiate between actual Neanderthal DNA and modern human or bacterial contamination. Recent advances in DNA sequencing technology made possible the actual determination of Neanderthal DNA genes and the comparison of these genes to known human genes. In this manner, the same technology that is enabling researchers to push forward the boundaries of modern medicine and agriculture is also helping to unveil the mysteries of our past.

http://www.BiotechPrimerBlog.com

CHAPTER THREE: LIFE'S BLUEPRINT

In this chapter, we have taken an in-depth look at the structure and organization of DNA. In the next chapter, we will examine how cells convert the information stored in DNA into the proteins that enable cells to do their job.

CHAPTER FOUR
From Gene To Protein

The process of converting the information contained in a gene into a protein is called **gene expression**. It is a multistep process in which cells build new proteins from chains of **amino acid** subunits. The order of amino acids is determined by the DNA sequence in a gene. There are two major steps involved in protein synthesis: **transcription** and **translation**.

During transcription, the original nucleotide sequence of the DNA code is rewritten into a messenger molecule called **messenger RNA (mRNA)**. RNA molecules are very similar to DNA but have a ribose sugar rather than deoxyribose, are single stranded, and use **uracil (U)** instead of thymine (T) as one of its four bases.

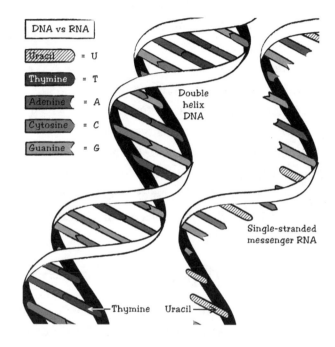

DNA vs RNA

Uracil = U
Thymine = T
Adenine = A
Cytosine = C
Guanine = G

Double helix DNA

Single-stranded messenger RNA

Thymine   Uracil

Each group of three nucleotides in an mRNA transcript, a codon, specifies a particular amino acid in the protein chain. There are 64 different potential codons corresponding to all the possible combinations of the four bases A, C, G, and U. Because there are only 20 amino acids, most amino acids are coded for by more than one codon. In the 1960's, scientists spent several years working out the genetic code—that is, determining what combinations of bases represent which amino acids. By 1966, the team led by Marshall Nirenberg at the National Institutes of Health had cracked the code. For this accomplishment, Nirenberg received the Nobel Prize for Physiology or Medicine in 1968. The results are displayed on the next page in what is referred to as a Codon Table.

> ## Cocktail Fodder
> When you fry an egg and see the egg white turn opaque as it cooks, you are observing the egg protein albumin come apart, or **denature**, which is what happens to proteins when they are exposed to excess heat.

GOING FURTHER

Why would nature create redundancy by allowing multiple codons to code for the same amino acid? This can be a protective measure against DNA mutations since a substitution mutation in the codon for leucine, for example, may yield a mutated codon that still codes for leucine. Recently, however, scientists have discovered that different codons coding for the same amino acid may be translated by the ribosome at different rates, which may influence the ultimate shape and function of the protein.

CODON TABLE

		Second Position									
		U		C		A		G			
		CODE	AMINO ACID	CODE	AMINO ACID	CODE	AMINO ACID	CODE	AMINO ACID		
First Position	**U**	UUU	phe	UCU	ser	UAU	tyr	UGU	cys	U	**Third Position**
		UUC		UCC		UAC		UGC		C	
		UUA	leu	UCA		UAA	STOP	UGA	STOP	A	
		UUG		UCG		UAG	STOP	UGG	trp	G	
	C	CUU	leu	CCU	pro	CAU	his	CGU	arg	U	
		CUC		CCC		CAC		CGC		C	
		CUA		CCA		CAA	gln	CGA		A	
		CUG		CCG		CAG		CGG		G	
	A	AUU	ile	ACU	thr	AAU	asn	AGU	ser	U	
		AUC		ACC		AAC		AGC		C	
		AUA		ACA		AAA	lys	AGA	arg	A	
		AUG	met	ACG		AAG		AGG		G	
	G	GUU	val	GCU	ala	GAU	asp	GGU	gly	U	
		GUC		GCC		GAC		GGC		C	
		GUA		GCA		GAA	glu	GGA		A	
		GUG		GCG		GAG		GGG		G	

To use this chart, start on the left hand side with the bold black letters. These letters represent the first position of the codon. The second position is at the top of the chart in bold. The third position of the codon is listed on the right hand side, again in bold. For clarity, each three letter codon is printed to the left of the amino acid it codes for.

The process of converting the mRNA into a protein is called translation. Translation involves a second type of RNA, **transfer RNA (tRNA)**, which picks up amino acids in the order specified by the mRNA codons and transfers them to the ribosome. The ribosome then forms a chemical bond—known as a **peptide bond**—between the two amino acids. As the ribosome moves along the RNA message, it continues to connect the specified amino acids to form an amino acid chain and the final protein molecule.

There are also nucleotide sequences that tell ribosomes where to begin and end translation. The **start codon** is AUG, and it codes for the amino acid methionine. As a result, the first amino acid of many proteins is methionine, although it is removed shortly after translation in some proteins. There are three **stop codons**: UGA, UAA, and UAG. The stop codons do not code for amino acids.

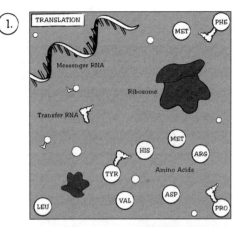

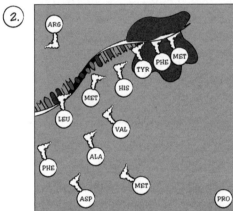

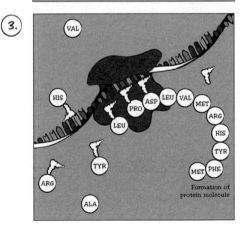

GOING FURTHER

A major goal within biotechnology is to understand, modify, and use proteins for a variety of different purposes. Because a protein's function is intrinsically connected to its structure, the ability to visualize protein structure becomes very important. In order to facilitate this understanding, scientists use computer programs to build three-dimensional molecular models of proteins.

There are two common ways in which these protein models are presented: space filling and ribbon. Space filling models are so named because the atoms that make up the protein are represented by spheres whose size are proportional to the size of the atoms and whose center-to-center distances are proportional to the distances between the atomic nuclei. Atoms of different chemical elements are usually represented by spheres of different colors.

Ribbon models show the overall path and organization of the protein backbone in three dimensions. The coiled ribbon in this representation represents a structural motif known as an alpha-helix—a right handed coil or spiral confirmation in the actual protein. The arrows represent beta-sheets—two or more parallel adjacent polypeptide chains.

1. Space Filling Model

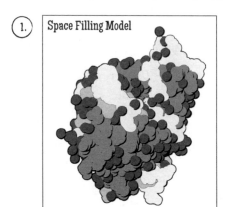

2. Ribbon Model

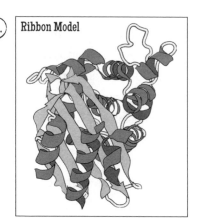

Proteins Take Shape

Ribosomes translate mRNA into a chain of amino acids like a string of beads, sometimes called a polypeptide chain. Each amino acid molecule has a common structure of three main parts—an amino group, an acidic carboxyl group, and an **R group**. The R group is what differs between each amino acid. The order, or sequence, of the amino acids in the chain determines the final structure of the protein molecule.

Some R groups attract one another, while others tend to repel one another, and it is this interaction between points along the chain, as well as with the aqueous environment of the cell, that causes a protein to fold into a particular shape. Each sequence of amino acids will form a characteristic protein structure, and it will form the same structure every single time it is translated. Moreover, no two protein structures are the same. Even though there are only 20 amino acids, the structure of each protein molecule is unique.

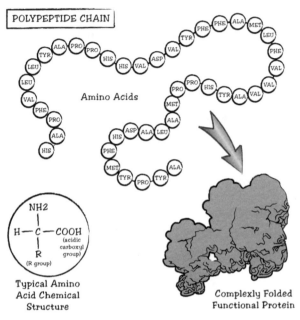

POLYPEPTIDE CHAIN

Amino Acids

NH2
|
H — C — COOH
| (acidic
R   carboxyl
(R group) group)

Typical Amino Acid Chemical Structure

Complexly Folded Functional Protein

Cocktail Fodder

Of the 20 different amino acids, ten are considered essential because they must be obtained from the diet; the other ten are termed nonessential because they can be made by the body. The essential amino acids are different for different organisms and even at different stages of life.

GOING FURTHER

Scientists are still trying to unravel the mysteries of how proteins fold. Over 100 types of protein folds have been discovered using technologies such as **x-ray crystallography** and **nuclear magnetic resonance**. The field of bioinformatics uses computer modeling to predict the three dimensional folds of new protein sequences based on comparison to the structures of known proteins. Understanding protein structure can assist scientists in identifying new drug targets and designing target-specific therapies.

Once an amino acid chain has folded to make a structure, it will often interact with other folded amino acid chains. In the picture shown here, four chains are interacting to form a structure that is different from the single chain. In fact, we can see that in forming this **tetramer**, the protein creates a new pocket that may function as an **active site**— an area of the protein that carries out its function.

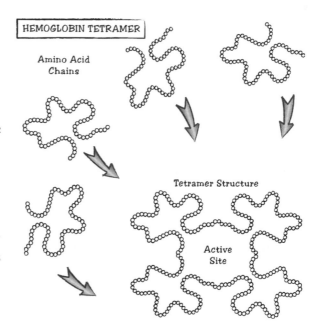

HEMOGLOBIN TETRAMER

Amino Acid Chains

Tetramer Structure

Active Site

Tricky Terminology

Meros is the Greek word for part. This is the origin of the root "mer", typically used in science to mean a repeating unit. The prefix in front of "mer" varies, depending on how many times the unit is repeated. A single unit is a monomer; two repeats, a dimer; three, a trimer; four, a tetramer. If the number of repeat units is large and/or variable, the term polymer is used, for "many parts".

INDUSTRY NOTE

When producing therapeutic proteins, scientists have to consider how to keep them stable so that they will function correctly when administered to the patient. Proteins are carefully engineered to make them as stable as possible, but most require refrigeration to maintain their folded structure and activity.

PRODUCTS AT WORK: Botox

A particularly well-known protein for sale is botulism toxin type A, marketed as Botox. Botox causes paralysis by binding to certain types of nerve cells and blocking signals for muscle contraction. First approved by the FDA in 1989 as a drug to stop uncontrolled blinking, eye ticks, excessive sweating, and other muscle spasms, Botox became approved as a cosmetic treatment three years later when scientists realized it also paralyzed facial muscles that cause wrinkles.

Protein Shape and Sickle Cell Disease

With an understanding of how gene sequence determines protein structure and function, we can now explore the effects of genetic variation on cell function, looking at sickle cell anemia as an example. Sickle cell anemia results from a single base **mutation**, which occurs in codon number six of the hemoglobin gene. The hemoglobin protein, coded for by the hemoglobin gene, is made up of four protein subunits—two alpha and two beta, which bind together to form the functional

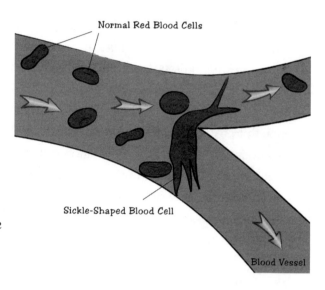

SICKLE CELL ANEMIA

Normal Red Blood Cells

Sickle-Shaped Blood Cell

Blood Vessel

hemoglobin **tetramer**. The sickle cell mutation occurs in the beta subunits, and this causes the formation of a "bump" in the protein's structure. This bump happens to fit into a notch in the beta subunit of another hemoglobin molecule. The bump on that beta subunit can fit into a notch on a third hemoglobin protein and so on, leading to the formation of rod-like hemoglobin chains.

Red blood cells containing normal hemoglobin are flexible and have a round donut shape. However, because the mutated hemoglobin gene produces hemoglobin molecules that form these rod-like structures, red blood cells containing this variant become rigid and sickle shaped. Sickle shaped cells do not pass through small blood vessels as easily as normal round cells and often get stuck, restricting blood supply to tissues and causing the complications of sickle cell disease. They are also removed from circulation and destroyed faster than normal red blood cells, causing anemia.

GOING FURTHER

Hemoglobin's job is to transport oxygen in red blood cells from the lungs to the body's tissues. The hemoglobin tetramer switches between two structural conformations, oxy and deoxy. In one conformation it binds oxygen, and in the other conformation it releases oxygen. This switch is sensitive to **pH**, or blood acidity. In the lungs, the pH is slightly higher than in the peripheral tissues, causing hemoglobin to take on its oxygen binding conformation. As blood leaves the lungs and enters tissues such as exercising muscle, the pH is slightly lower, or more acidic. This slight change in pH is enough to cause a conformational change in the hemoglobin protein, and oxygen is released. The conformation switch is essential for hemoglobin's ability to pick up oxygen in the lungs and release it in the tissues.

The effect of the sickle cell mutation also explains the increased malaria resistance associated with sickle cell trait. The degree of red cell sickling depends on oxygen tension; the lower the oxygen content, the more sickling. The malaria parasite *Plasmodium falciparum* spends part of its life cycle in red blood cells. Since *Plasmodium* uses up oxygen for respiration, the parasite decreases the oxygen content of the infected red blood cells, increasing sickling, thereby increasing removal and destruction of infected cells. By this mechanism, sickle cell trait eliminates the parasites.

Cocktail Fodder

Because proteins are so easily broken apart by the body's digestive system, patients must receive therapeutic proteins as an injection.

Protein Levels and Cancer

Differences in gene expression—how much protein is made—can sometimes give clues to the root causes of disease. Approximately 25% of breast tumors are HER2-positive—they express high levels of a gene called **HER2** that codes for a growth factor receptor in the cell membrane. Its function in normal breast tissue is to stimulate cell growth in mammary ducts and glands. The HER2 protein is usually expressed at very low levels and only stimulates cell growth in response to growth factors. When overexpressed, it is responsible for the uncontrolled growth of tumor cells and their resistance to cell death. Whether or not breast tumors overexpress HER2 has a big influence on the effectiveness of anticancer drugs; HER2-negative tumors respond better to some drugs, and HER2-positive tumors respond better to others. Measuring a patient's HER2 expression level helps to determine the most suitable cancer treatment. There are commercially available diagnostic kits that enable doctors to measure HER2 expression levels directly from a biopsy sample.

PRODUCT AT WORK: Companion Diagnostics

HER2-positive breast cancers are usually very aggressive, and HER2 overexpression is associated with poor prognosis. Therefore, HER2 is regarded as an important target for anticancer drugs, some of which are already commercially available. After an initial breast cancer diagnosis, a physician will typically request further diagnostic analysis of the biopsy sample to determine which subtype of breast cancer is present, and prescribe the best treatment possible for that subtype. A diagnostic used to inform a physician's prescribing decision in this way is called a **companion diagnostic**.

RNA Interference

RNA interference, or RNAi, is a gene silencing technique that was discovered in the 1990's to occur naturally in petunia plants. The basic idea is this: mRNA in cells is normally single stranded. Cells contain enzymes called Dicer and RISC that recognize double stranded RNA and destroy it. This is in part because the presence of double stranded RNA could signify a viral infection, so their destruction may have evolved as a type of cellular defense against viruses.

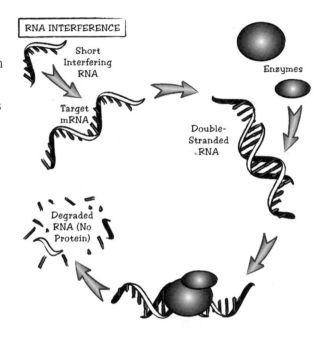

Researchers can take advantage of these naturally occurring enzymes to "silence", or block the expression of, genes of their choosing. These genes are referred to as target genes. This can be done by designing short pieces of RNA to hybridize to the complementary sequence of the mRNA encoded for by the target gene. This short interfering RNA—RNAi—can be synthesized in the lab and transferred into cells, where it will hybridize to the target mRNA molecules, forming a double stranded RNA molecule. Dicer and RISC then recognize the double stranded RNA and degrade it. Without an RNA message, no protein is made.

INDUSTRY NOTE

RNAi has had much success in basic research labs, helping investigators to tease out the functions of specific genes and their roles in various diseases by blocking their expression and then measuring how the cell is impacted. Many different biotech companies are working to develop drug therapies based on this technology. In theory, it would be a highly specific—and highly effective—way to block the production of certain disease causing proteins. In practice, however, it has been difficult to transfer this technology into therapeutic use because of the challenges in getting the therapeutic RNAs to the right location within the human body, in high enough concentrations to be effective at gene silencing. Scientists are working on developing RNAi molecules with chemical modifications to make them more stable, as well as on developing better methods of delivery.

GOING FURTHER

Certain tissues of the body make better targets for RNA therapies, because the RNAi has been shown to accumulate there. These tissue targets include the liver, kidney, spleen, bone marrow, and fat cells. Thus diseases that affect these tissues are good candidates for treatment with RNAi. Eyes and lungs are also considered to be good candidate tissues, because of the relative ease of drug delivery using eye drops or inhalants, respectively.

miRNA as a Diagnostic?

Over the past ten years, **micro RNAs** (miRNAs) have gained attention on a number of fronts. What exactly is miRNA? miRNAs are short pieces of RNA that do not code for proteins. Rather, they serve as a cellular mechanism for the regulation of gene expression. They do this by actually interfering with the translation—or the conversion into a protein—of "target" mRNA molecules in the cell. miRNAs are very closely related to RNAi molecules.

Scientists involved in basic research are studying miRNAs in order to better understand gene expression and regulation in organisms at all stages of development, as well as in disease vs. non-disease states. Scientists interested in drug development have worked on developing therapeutics based on miRNAs—designing miRNAs that specifically interfere with the production of a disease-associated protein. Although this approach shows much promise in the lab and in animal studies, it has not yet been successfully translated into humans.

The latest development on the miRNA front is the observation that these molecules may have the potential to be developed into a unique type of diagnostic. Several different groups involved in researching the molecular pathogenesis of various types of cancer have all come to the same conclusion: in a wide range of cancers, including gastric, colorectal, pancreatic, and liver, there are significant variations in the amounts and types of these miRNA molecules, as compared to normal tissues. This means that these aberrant miRNAs could serve as a biomarker for cancer—perhaps long before any other symptoms are present, making them a potential "early warning signal" that is so important for increasing the likelihood of a good outcome for cancer patients.

Excerpt From Biotech Primer Blog Continued From Previous Page

As with the use of miRNAs as a therapeutic, it remains to be seen whether or not these observations will translate well into the clinic. But hopes are high that they may herald the next generation of highly specific, relatively non-invasive, early detection method for cancer diagnosis.

http://www.BiotechPrimerBlog.com

In this chapter, we described the process that cells use to convert the information in genes into proteins. In the next chapter, we will examine how genetic variation—differences in gene sequences between individuals—contributes to our understanding of disease and the development of new therapeutics and diagnostics.

CHAPTER FIVE
Mutations: The Spice Of Life

Our genomes—everyone's genomes—are about 99.9% identical. This is why we can speak of something called "the human genome". However, in a genome that contains 3 billion base pairs, even a 0.1% difference in sequence can lead to significant variation within the human population. This genetic variation accounts for the differences we observe between individuals— trivial differences such as distinct in eye, hair, and skin color; more profound but not necessarily harmful differences such as variant blood groups A, B, AB, and O; and some medically significant differences such as susceptibilities to various diseases and responses to distinct drugs.

Genetic diversity arose as a result of **mutations** in the DNA sequence.

> **Mutations can occur by different means, and the outcome for the organism can vary.**

Mutations can occur by different means, and the outcome for the organism can vary. A **substitution**, or **point mutation,** results when one base is swapped out for another. If the DNA polymerase enzyme accidentally places a C where a G should be during DNA replication, the substitution yields a different codon. Sometimes DNA polymerase may skip over a base, which is a **deletion**, or add an extra base to the sequence, which is an **insertion**. If a mutation goes unrepaired, it results in DNA sequence changes that will then be copied, becoming permanent. Changes in the DNA sequence as a result of errors by the DNA polymerase during DNA replication are rare, but they do happen. This is understandable when you consider the fact that every time a cell divides, it must copy all 3 billion base pairs of DNA in just a few hours! Fortunately, the DNA polymerase also has proofreading ability, so it is able to "catch" and correct most mistakes, but occasionally a mutation goes unrepaired, resulting in DNA sequence changes that will then be copied, becoming permanent. In addition to errors made during DNA replication, mutations can also result from environmental factors, such as radiation from the sun or x-rays or from chemicals in cigarette smoke.

There are a number of mutations in each and every one of your cells. This is not necessarily cause for alarm—not all mutations are harmful. Mutations can have various effects on the function of a gene and its protein product. Some, of course, are deleterious, damaging the function of the protein encoded for by the mutated gene. However, many mutations turn out to be neutral and

Tricky Terminology

Original sequence:
ATGACTGCATGTTACGGT

Substitution mutation:
ATGACTGCA**C**GTTACGGT

Deletion mutation:
ATGACTGCA - GTTACGGT

Insertion mutation:
ATGACTGCA**GC**TGTTACGGT

have no effect on gene function. How is this possible? Recall from the last chapter that only 1.5% of the human genome actually codes for proteins—the rest of it is noncoding, formerly known as junk DNA. So odds are, most of the time that a mutation occurs, it will occur in these noncoding regions, and not affect a protein's structure or function. The other reason mutations may be neutral lies in the redundancy of the genetic code—the fact that most amino acids are coded for by more than one codon, as described in the last chapter. So in some cases, even if a mutation occurs in a protein coding region of the genome, the same protein gets made. Very rarely, mutations can be adaptive and have a beneficial effect on gene function, conferring an advantage on the organism. For example, a slight change in an enzyme's structure due to a mutation may cause it to function more efficiently. This is the basis of Charles Darwin's **Natural Selection** or "survival of the fittest" theory. **Evolution** is the natural selection of beneficial changes.

GOING FURTHER

All dogs have the same DNA but with slight variations that allow breed characteristics, like short legs or floppy ears. Dog breeding is an example of unnatural or artificial selection—the selection was done by people, not nature.

Some mutations are passed on from one generation to another, and some arise during an organism's life span. Most mutations occur in **somatic cells**; that is, cells that are not **gametes** (eggs or sperm). These mutations are called somatic mutations and are not passed on to children. But mutations that do occur in sperm or egg cells, germ line mutations, will be inherited. If the mutation is so severe that an organism cannot survive, that organism does not pass it on to the next generation. Therefore, over time, deleterious mutations exit the gene pool.

ARTIFICIAL SELECTION

Single Nucleotide Polymorphism

The type of mutation we called a substitution earlier, where one base is substituted for another base, is more commonly known in industry as a **single nucleotide polymorphism** (**SNP**, pronounced "snip"). The small differences between different genomes (base differences) are called **polymorphisms**—

> ## Tricky Terminology
> The term somatic derives from the Greek word soma—of the body. Thus somatic cell is the term used to refer to any cell of the body other than the gametes (egg and sperm). Gamete derives from the Greek words for husband (gametes) and wife (gamete). Gametes are also referred to as **germ cells**.

generally speaking, if a genetic variation occurs in more than 1% of the population, it's called a polymorphism. SNP's are by far the most common type of polymorphism, and they are most often the kind of variation that contributes to the most obvious differences between us. However, other kinds of polymorphisms also have a big impact on genetic diversity. Examples include multiple base insertions or deletions. In some cases, there are even differences in the number of copies of a particular gene— referred to as copy number variation.

A fairly common SNP accounts for why people have different responses to bitter tasting food. Humans can discern five different tastes: sweet, sour, salt, bitter, and unami (the taste of MSG). The food chemicals responsible for tastes—like sugar and artificial sweeteners for sweetness, or acetic and citric acids for sourness—attach to receptor proteins located on the outside of our taste buds. When a food chemical like citric acid attaches to the taste receptor, a nerve signal travels from the taste bud to the part of the brain that processes taste sensation. The codes for the taste receptor proteins are kept by the taste receptor genes; the sweet taste receptor gene, the unami taste receptor gene, and so on.

A SNP in the bitter taste receptor, called TAS2R, affects the way the receptor responds to bitter tasting chemicals and sends the signal to the brain. People with one form of the SNP therefore respond differently than people with the other form of the SNP. Thus the same food can actually taste different to two different people depending on their particular taste receptors.

Cocktail Fodder

Barbara McClintock won a Nobel Prize in Physiology or Medicine in 1983 for showing that genes can move or "jump" to a new place on a chromosome. She did this by studying a type of multicolored corn known as maize. The dark and yellow kernels are caused by moving genes, which is a type of mutation called a translocation.

Cocktail Fodder

Scientists estimate that, between individual human beings, there is a single base difference per every 3000–5000 bases, which is what makes us different from each other.

GOING FURTHER

We just learned that humans are 99.9% genetically similar. Let's take a look at one way we may differ. Phenylthiocarbamide, also known as PTC, or phenylthiourea, is an organic molecule. It has the unusual property of either tasting very bitter, or being virtually tasteless, depending on the genetic makeup of the taster.

The ability to taste PTC is a dominant genetic trait. The test to determine PTC sensitivity is one of the most common genetic tests in humans, even having been used to test paternity before more sophisticated DNA testing was available. About 70% of people can taste PTC, varying from a low of 58% for Aboriginal people of Australia and New Guinea to 98% for Indigenous peoples of the Americas. One study has found that populations of non-smokers and those not habituated to coffee or tea have a statistically higher percentage of PTC tasters than the general population.

People who find PTC bitter may also find cigarette smoke to be more offensive, and it may also help explain the vast gulf between the people who enjoy coriander and those who find it unpalatable.

Genetic Basis of Disease

Mutations play a large part in disease. In a **monogenic disease**, changes in one gene cause the disease. Examples of monogenic diseases include sickle cell anemia (as described in Chapter Four), cystic fibrosis, and Huntington's disease. In contrast, **polygenic diseases** are caused by the interactions of many different genes. Polygenic diseases are more common than monogenic diseases, and include cancer, heart disease, Alzheimer's disease, and Parkinson's disease. Polygenic diseases often have **susceptibility genes** associated with them that increase the likelihood of the person developing the disease, but do not absolutely predict its development—the ultimate disease outcome will depend on various other genes in the individual's genome, as well as environmental factors. An example of susceptibility genes is the association of breast cancer with the BRCA1 and BRCA2 genes.

INDUSTRY NOTE

As a polygenic disease, Alzheimer's presents a particular challenge, and also an opportunity, for the biopharmaceutical industry. People over the age of 85 have a 50% chance of getting the disease, but researchers do not yet understand what causes it or how to effectively treat it. Some existing medications may only work in about 20% of Alzheimer's patients, and even then, they may not work for longer than two years.

Polygenic Disease: Cancer

Cancer can be described as uncontrolled cell growth. Healthy cells have tight controls on cell division, and only divide in response to outside signals promoting cellular growth and division. Noncancerous cells also respond to signals that tell them to *stop* dividing; for example, most exhibit contact inhibition, meaning that if they touch a neighboring cell, they stop dividing. Cancer cells have lost many of these controls or checks on cell division and start dividing and continue to divide inappropriately.

What causes this loss of control? Typically, it is the result of mutations in one or more of the genes that code for proteins whose primary function it is to regulate cell division. For example, a growth factor receptor is a protein that signals the cell to grow and divide in response to chemical signals called growth factors. If a mutation occurs in this type of protein, it could signal a cell to divide even in the absence of growth factors. Other types of proteins respond to signals that tell the cell to stop dividing; mutations in these proteins could cause the cell to ignore the "stop" signals. In most cases, a cancerous tumor will have mutations in several different genes involved in the regulation of cell division.

GOING FURTHER

Why does cancer make people so sick? The answer varies depending upon the type of cancer, but all cancers share the trait of out-of-control cell growth. This excessive growth puts extreme stress on the tissue that it affects. Cancerous cells—tumors—no longer carry out the function of the tissue it is growing in, yet they take up space and absorb nutrients, preventing the healthy parts of the tissue from functioning correctly. Tumor cells steal resources from healthy tissue through a process referred to as **angiogenesis**. During angiogenesis, the tumor sends out chemical signals to neighboring blood vessels, rerouting them to the tumor itself. In this way, the cancer cells get the extra nutrients they need to support their growth, as well as a mechanism for removing the additional metabolic waste products generated. Multiple companies have developed drugs to interfere with the process of angiogenesis.

The **p53** gene is the most frequently mutated gene in human cancer, and the p53 protein it produces is often called the "guardian of the genome". Its role is to make a decision whether to repair DNA or to kill the cell in response to DNA damage. To perform this function, p53 has to bind to DNA in a very specific manner.

Image 1 shows p53 bound to DNA. Four domains of the protein bind to DNA in a cooperative manner. Image 1 also shows one of the domains bound to DNA where the DNA binding surface of the p53 molecule fits into the grooves of the DNA helix.

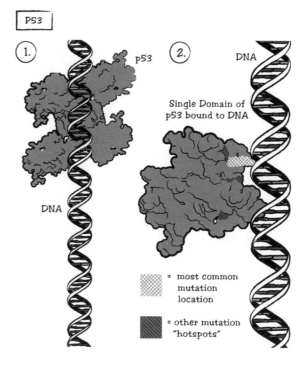

Most of the mutations that occur in human cancer—mutation "hotspots" shown on image 2 in a striped pattern—occur in the DNA binding domains, close to the DNA binding surface. The most common mutation shown in checkered pattern, also on image 2, replaces an arginine residue, which snakes into the DNA groove. Changing this amino acid to almost any other one is enough to disrupt binding of p53 to DNA and disrupt p53's guardian duties.

Research into how protein structure abnormalities cause cancer and other diseases is leading to the discovery of new drugs, some of which are designed to target the aberrant protein structures.

Monogenic Disease: Sickle Cell Anemia

As discussed earlier, people inherit two copies of each gene. One copy comes from the mother, the other from the father, and each copy is known as an **allele**. Most disease genes are recessive, meaning that they only cause the disease if the disease form of the gene is inherited from both the mother and the father. A person would have to have two disease alleles in order to have a recessive disease. People with one disease allele are said to be **carriers**. They don't have the disease, or their symptoms are very mild, but they can pass on the disease if their spouse is also a carrier.

Sickle cell anemia, for example, results when someone has two hemoglobin alleles with the sickle cell SNP. A person with one copy of the sickle cell SNP will have sickle cell trait. This is a much milder form of the disease with relatively few symptoms. However, people with sickle cell trait are resistant to infection by the parasite that causes malaria, which is very common in some parts of the world, particularly Africa. Since they have protection against malaria, one of the world's deadliest diseases, people with one copy of the sickle SNP have a survival advantage over those with two normal hemoglobin alleles in areas with a high incidence of malaria. The sickle cell SNP has therefore been selected for in most of Africa. The incidence of sickle cell disease among Africans is about 4%, while among African Americans, who are not frequently exposed to malaria carrying mosquitoes, it's 0.25%. This is an example of natural selection of a polymorphism in populations.

There are many other examples of selection for, or against, a particular polymorphism in certain populations. For instance, a polymorphism in the **CCR5 gene**, which is prevalent in people of European ancestry, can offer some protection against HIV infection. Scientists believe that the CCR5 polymorphism may also provide protection against some other infectious diseases and that it was selected for during the bubonic plague epidemics—the Black Death—that swept Medieval Europe roughly one thousand years ago.

International HapMap Project

Researchers have launched a project to map all of the human SNPs. The goal is to match people's SNP profiles with their susceptibility to diseases. However, there are about ten million different SNPs in the human genome. To narrow down analysis of all those SNPs, a collection of government and privately funded teams from different countries started the **International HapMap** (haplotype map) **Project** in 2002.

SNPs tend to be inherited in blocks called **haplotypes**. Each of these haplotypes can be identified by one or two representative SNPs, called **tag SNPs**. The HapMap Project will reduce the number of SNPs to analyze from ten million SNPs to about half a million tag SNPs. In the initial phase of the project, the genomes of 270 people from four different geographical populations were analyzed to start building a representative HapMap. The HapMap is being used to compare haplotypes between different groups of people—for instance, between people with or without diabetes—to try and identify SNPs that are associated with diabetes. The HapMap is already proving to be an invaluable resource for researchers using genome-wide analysis to hunt for disease susceptibility genes. The ultimate aim is to construct a database to house the genetic information that will predict an individual's susceptibility to certain diseases or the way that person is likely to respond to drugs, vaccines, or environmental factors.

Cocktail Fodder

In 1996, the biopharmaceutical company deCODE genetics, Inc. of Iceland wanted to undertake the building of an Icelandic Health Sector Database (HSD), that would contain the medical records and genealogical and genetic data of all Icelanders. Iceland's Supreme Court shut the project down in 2003 due to issues of privacy and consent.

Another important, but sometimes overlooked, aspect of the HapMap Project is the ethical issues it raises. The anonymity of all DNA donors to the HapMap genome database is highly protected. However, genetic information about certain populations of people, based on geographical and cultural determinants, will become available. For example, the data will reveal if there is a higher prevalence of a disease susceptibility SNP in one population group compared to another, which could lead to the presumption that each member of that group has a higher risk of disease. This kind of information will inevitably surface, and the HapMap Project is trying to counter possible misinterpretation of the data by informing and educating.

At the outset of the HapMap Project, some people feared that populationwide haplotyping could reinforce stereotypical ideas of race. In fact, the reverse has been true, and the HapMap Project clearly shows that concepts of race have little or no basis in genetics and are largely the consequence of cultural and social background.

INDUSTRY NOTE

If a drug carries a significant risk of serious adverse events, the FDA requires the drug maker to include a boxed warning (also referred to as a black box warning) describing the risk on the product labeling or package insert.

Pharmacogenomics

The understanding that most diseases have genetic influences has led to a new area of medicine called **pharmacogenomics**. The goal of pharmacogenomics is to use the information contained within a patient's genome—a sort of genomic signature—to design the best treatment plan for that patient. It is often referred to as personalized medicine, and it takes into account not only potential disease genes, but also genes that might affect how the patient responds to a particular drug.

CHAPTER FIVE: MUTATIONS: THE SPICE OF LIFE

The use of pharmacogenomics has the potential to positively impact medicine in several ways. Researchers and companies can use pharmacogenomics to predict drug effectiveness. Pharmacogenomics will allow a physician to select the drug most likely to be beneficial for a particular patient or, from another perspective, to select the patients most likely to respond to a given drug. It can also be used to predict drug safety. Approximately 100,000 people die each year from adverse drug reactions, making it one of the leading causes of death in the U.S.. The reason some people experience adverse side effects, while most do not, is largely because of genetic differences, particularly in the genes that metabolize, or break down, drugs. Using pharmacogenomics, a physician can avoid prescribing unsafe drugs for a particular patient. Similarly, scientists can use analysis of drug effectiveness and safety prediction in combination to formulate an appropriate dosing schedule for the patient. This is important not only in safely and effectively treating disease, but also in controlling drug costs.

Designing drugs to target a particular genotype will allow fine tuning of the drug development process from the starting point. Clinical trials will be improved and streamlined by selecting patients for trials based on their genomic signature, resulting in shorter, less expensive trials. Drugs that have failed trials because of lack of effectiveness might be reevaluated in genomically selected patient populations to see if they will be effective for a particular genomic niche. As an example, the lung cancer drug Iressa did not prove effective in treating patients until scientists realized it only works in a subset of patients with a mutation in a gene called EGFR.

Biotechnology companies may be able to reduce drug costs to patient and provider by targeting only relevant genotypes, selecting appropriate clinical trial protocols, and focusing marketing on relevant patient markets. Finally, screening for genes that predispose to disease will give people the information they need to take appropriate precautionary measures against developing disease. This kind of screening is underway for some diseases, such as breast cancer. Women from families with a history of breast cancer can now screen themselves for the breast cancer susceptibility genes BRCA1 and BRCA2.

Personalized Medicine in Practice

Personalized medicine is already playing a role in some areas of healthcare. In this section we will explore two examples, starting with a look at some of the genes involved in cancer.

Genentech designed a monoclonal antibody drug called Herceptin to be effective against breast cancer cells that produce very high levels of the **HER2** protein, a growth factor involved in promoting cell division. These cancer cells are referred to as HER2-positive cells. Since Herceptin does not work in patients with HER2-negative tumors, patients can be prescreened to predict the likely effectiveness of Herceptin. More recently, scientists have discovered that

PERSONALIZED MEDICINE: FINDING THE RIGHT THERAPEUTIC

HER2-
will not respond
to Herceptin

HER2+
some will respond
to Herceptin

HER2+ / PTEN+
most will respond
to Herceptin

the status of at least one other gene can predict Herceptin effectiveness. **PTEN** is the second most common tumor suppressor gene mutated in human cancer (**p53** is the most common). The normal function of tumor suppressor genes is to protect cells from becoming cancerous. When PTEN is mutated, this function fails, and cells are more likely to become cancerous. Mutation of PTEN in HER2-positive breast tumors makes Herceptin far less effective in stopping the cancer's growth. Therefore, patients who are HER2-positive and do not have mutations in PTEN (PTEN-positive) are most likely to respond well to Herceptin. It is very likely that other genes will impact the effectiveness of Herceptin, as well as other cancer drugs, and that we will become more efficient in predicting cancer drug response.

Tricky Terminology

The presence of a functional copy of a particular gene is typically indicated with a '+'. So breast cells that contain a copy of the tumor suppressor PTEN gene and produce adequate amounts of functional PTEN protein, are referred to as $PTEN^+$; cells that lack the gene, or do not produce functional PTEN protein, are referred to as $PTEN^-$.

In the case of $HER2^+$ cells, the designation means that these cells produce excessive amounts of the HER2 receptor. Breast cells that are not designated $HER2^+$ still produce HER2 protein, but at much lower levels. Tricky terminology indeed!

Another example of applications in personalized medicine involves a large family of genes called cytochrome P450, or CYPs. These genes code for proteins that work to detoxify chemicals in the liver. They also metabolize many prescription drugs by breaking down the active drug into inactive products. SNPs often occur in CYP genes, and this can sometimes affect the ability of the CYP proteins to metabolize drugs. A SNP that decreases the activity of a CYP will slow down metabolism of the drug and cause unexpectedly high levels of drug in the bloodstream for longer than

Cocktail Fodder

Some naturally occurring chemicals can decrease the ability of CYPs to metabolize drugs. For instance, grapefruits contain a chemical that blocks the activity of CYP3A4, which is why many drug instruction labels advise against drinking grapefruit juice when taking certain medications.

expected periods. This can have very serious implications for drug effectiveness and safety. Many serious adverse reactions to antidepressant drugs, for instance, have been attributed to SNPs in CYP2D6. Therefore, knowing the CYP genotype of a patient will be very important for designing an appropriate drug dosing schedule.

PRODUCTS AT WORK: AmpliChip

Cytochrome P450 gene screening represents the first commercially available pharmacogenomic screening kit. Roche sells the AmpliChip, a DNA array or gene chip, that allows screening for SNPs in selected CYPs.

EXCERPT FROM BIOTECH PRIMER BLOG

Personalized Medicine: Seasonal Variations

A class of liver enzymes known as cytochrome P450 (CYP450) enzymes are largely responsible for breaking down most prescription drugs. Researchers and physicians are already aware of specific genetic differences that make some individuals metabolize certain drugs more quickly or more slowly than other individuals, and a number of companies are developing diagnostic tests to identify which type of metabolizer a particular patient is: poor, normal, or ultra-metabolizer. Different drugs are metabolized by different subcategories within this broad CYP450 family, so a patient may be a poor metabolizer for one type of drug, and a normal metabolizer for another type.

A recent publication in the journal *Drug Metabolism and Disposition* indicates that a person's ability to metabolize certain drugs may also be influenced by how much sunlight they get. This is because one subcategory of the CYP450 enzymes, referred to as CYP3A4, is activated by higher levels of vitamin D, which in turn is produced by a person's skin in response to sun exposure. A research team led by Erik Eliasson of the Karolinska Institute found that the levels of certain drugs known to be broken down by the CYP3A4 enzymes varied significantly, depending on the time of year and relative amounts of sunlight exposure, with higher drug levels corresponding to the winter months when the CYP3A4 enzymes were less active. This pattern was not seen in drug levels known to be metabolized by different CYP enzymes.

Of course, this seasonal variation will have the most profound impact in places where differences in sunlight exposure vary significantly throughout the year. It's not a coincidence that the study's authors live in Sweden, where the average number of daylight hours in December is six and one-half, and

Excerpt From Biotech Primer Blog Continued From Previous Page
the average number in June is eighteen. In most regions of the world, the variation is not so extreme; still, it does exist, and it will be interesting to see if these less pronounced variations also influence drug metabolism. Physicians may indeed have yet another variable to take into account in order to determine the most accurate drug dosages.

http://www.BiotechPrimerBlog.com

In this chapter, we have discussed genetic variation and its impact on medicine. Next we will describe the process by which scientists transfer genes from one organism to another—genetic engineering. It is this technology, genetic engineering, that created an industry.

CHAPTER SIX

The Technology That Created An Industry

Genetic engineering technology allows scientists to introduce new genes into organisms. Since the DNA of every living thing consists of the same building blocks, the DNA of different organisms is compatible, and researchers can join DNA pieces from different organisms to create **recombinant DNA**. Genetic engineers insert new genes into crops to give the crops desirable traits, such as resistance to pests. They also insert human genes into other organisms to produce human proteins.

Cocktail Fodder

The introduction of genetic information from spiders into goats allows these mammals to produce spider silk in their milk, which is collected and purified to make products such as soft body armor.

As we learned earlier, researchers developed the first biotech therapeutic, human insulin, using recombinant DNA in bacteria.

Transgenic Organisms

A **genetically modified organism (GMO)** or **genetically enhanced organism (GEO)** is an organism whose genetic material has been altered using genetic engineering techniques. These techniques, generally known as recombinant DNA technology, use DNA molecules from different sources that are combined into one molecule to create a new set of genes. Scientists transfer the new DNA into an organism, giving it modified or novel genes.

Transgenic organisms, a subset of GMOs, are organisms that have inserted DNA originating from different species. The ability to introduce human genes into model organisms such as mice is a research tool that enhances our understanding of diseases and facilitates drug discovery. Transgenic technology has also had major impacts in the production and nutritive value of food crops and livestock. For instance, researchers have recently engineered carrots to express a gene called sCAX that causes the carrots to store much higher levels of calcium than regular carrots. Scientists found in preliminary tests that people eating these carrots had increased calcium absorption, or bioavailablity, by an average of 41%. This could be significant in the prevention of osteoporosis, a bone disease caused, in part, by low calcium uptake from diet.

Cocktail Fodder

In 2000, the artist Eduardo Kac introduced the world to an artistic creation named "Alba", a genetically engineered rabbit that glows in ultraviolet light. Scientists produced Alba by giving her a gene from a glowing jellyfish that makes her skin cells produce green fluorescent protein.

Uses of Transgenic Organisms

Scientists have found many other uses of transgenic technologies in addition to creating glowing rabbits and calcium enhanced carrots. In this section, we'll take a look at some of the various applications using transgenic organisms to improve human health and nutrition. The three general categories of transgenic organisms are microbes, plants, and animals.

The first transgenic organisms were *E. coli* bacteria **transformed** with recombinant DNA by Stanley Cohen and Herbert Boyer, using technology that enabled the founding of Genentech in 1976. The production of transgenic bacteria, in a sense, marks the birth of biotech. Transgenic bacteria have since been used for the production of a whole range of therapeutic human proteins including human insulin and human growth hormone.

Tricky Terminology

The term "recombinant" may sound highly technical, but it really just refers to the process of combining DNA from two different sources to produce a new DNA molecule.

INDUSTRY NOTE

Researchers are developing ways to efficiently produce the antimalarial drug Artemisinin using genetically engineered yeast. Current production techniques require isolating the drug from wormwood plants—an expensive process resulting in an average cost of $10 per dose—far too expensive for most developing countries. Optimizing production in yeast could bring the cost per dose down to as little as fifty cents.

Scientists have found environmental applications of transgenic microbes, too. Bacteria can play a role in the degradation of oil spills, the removal of heavy metals from water or soil, and the production of fuel —so called microdiesel created by genetically engineered bacteria from bulk plant material.

Transgenic plants are being used for the production of therapeutic proteins and vaccines, and to create better crops. Genetically modified (GM) plants account for significant percentages of food and textile crops, particularly corn, soybean, cotton, and canola. The U.S. is their major producer, but Argentina, Brazil, Canada, China, and India are also growing significant amounts of GM crops. Genetic modifications improve resistance to pests, herbicides, pesticides, viruses, and harsh environmental conditions, and also improve nutritive value.

Genetic engineering in animals has produced transgenic mice that are essential models for understanding disease mechanisms. Proteins of medical or economic value can be produced in renewable sources from transgenic animals—for example, biologic drugs in cow's milk, spider silk in goat's milk, and high yield wool from sheep. Researchers are also working to increase the nutritive value of livestock, introducing genetic modifications that will produce leaner cows and pigs that contain omega-3 fatty acids. The techniques used to create genetically modified animals will be described at the end of this chapter.

GOING FURTHER

The use of animal organs in people is called **xenotransplantation**, and the major problem with using organs from animals for organ transplantation is that of hyperacute tissue rejection. The patient's immune system immediately recognizes the animal organ as foreign and attacks within minutes. However, by introducing human "identity" genes into pigs (pig organ size is similar to human), researchers hope to cause the expression of human proteins on the outside of the transplanted organ, thereby instructing the patient's immune system that the organ is human.

Recombinant DNA

In the early 1970's, scientists discovered a type of enzyme in bacteria called a restriction enzyme that could cut, like molecular scissors, through DNA strands at particular base sequences.

The illustrations to the right show how restriction enzymes work. Different restriction enzymes recognize and cut different, but specific, DNA sequences. For example EcoR1 always recognizes the sequence GAATTC and always cuts between G and A (G AATTC) as shown at the bottom of the illustration. Scientists will choose which restriction enzyme to use based on where they want to cut the DNA strand.

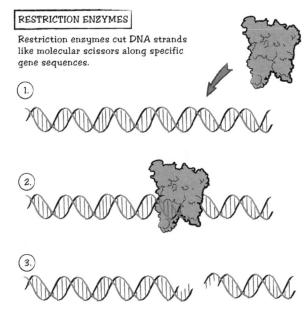

RESTRICTION ENZYMES

Restriction enzymes cut DNA strands like molecular scissors along specific gene sequences.

1.

2.

3.

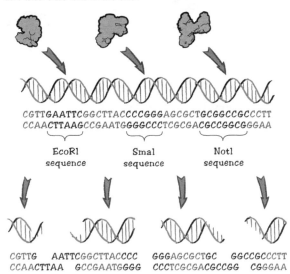

Each restriction enzyme recognizes a specific code sequence and then cuts within the site:

CGTTGAATTCGGCTTACCCCGGGAGCGCTGCGGCCGCCCTT
CCAACTTAAGCCGAATGGGGCCCTCGCGACGCCGGCGGGAA

EcoR1 sequence	Smal sequence	Notl sequence

CGTTG   AATTCGGCTTACCCC   GGGAGCGCTGC   GGCCGCCCTT
CCAACTTAA   GCCGAATGGGG   CCCTCGCGACGCCGG   CGGGAA

These restriction enzymes enable researchers to cut DNA from two different species—for instance, from bacteria and human cells—and glue them back together. The glue they use is DNA ligase, also an enzyme from bacteria, and the hybrid DNA molecule is referred to as recombinant DNA.

The type of bacterial DNA most often used by scientists to make recombinant DNA is called a **plasmid**. A plasmid is a small, circular DNA molecule that is capable of replicating independently from the chromosomal DNA. Plasmids occur naturally in bacteria, where they increase genetic diversity because they can be transferred from microbe to microbe. Scientists use plasmids as **vectors** to carry a particular gene of interest into bacteria cells.

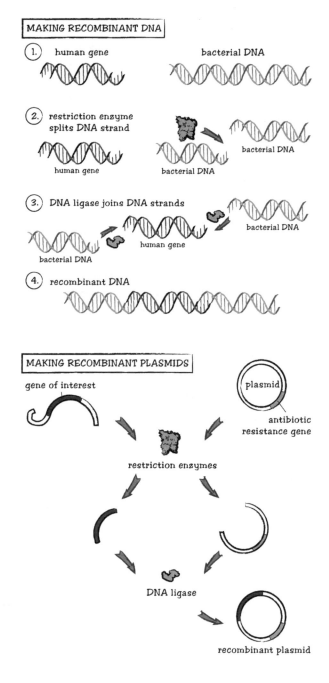

MAKING RECOMBINANT DNA

1. human gene   bacterial DNA

2. restriction enzyme splits DNA strand
   human gene   bacterial DNA   bacterial DNA

3. DNA ligase joins DNA strands
   bacterial DNA   human gene   bacterial DNA

4. recombinant DNA

MAKING RECOMBINANT PLASMIDS

gene of interest   plasmid
                    antibiotic resistance gene

restriction enzymes

DNA ligase

recombinant plasmid

Engineering Bacteria

Once a researcher creates a plasmid with a particular gene of interest, he or she can integrate it into the bacteria through a process called **transformation**. Transformation is the genetic alteration of a cell resulting from the uptake, genomic incorporation, and expression of foreign genetic material. Transformation can be carried out in the lab through chemical methods that change the permeability of the cell membrane or by **electroporation**, which creates small transient holes in the cell membrane after electrical shock.

Tricky Terminology

When a recombinant plasmid is transferred into bacterial cells, the cells are said to be **transformed**. When a recombinant plasmid is transferred into other cell types, the cells are said to be **transfected**.

The process of transformation is never 100% efficient—there will always be some bacteria that fail to take up the plasmid. Since some bacteria divide as often as once every 20 minutes, it is important to immediately establish a culture with only bacteria containing the introduced plasmid. Otherwise, the recombinant bacteria may be lost by growth competition from nonplasmid containing bacteria. As illustrated on the following page, the way scientists exclude bacteria that do not take up the plasmid is to include an antibiotic resistance gene in the plasmid. This is a gene that produces a protein that make the bacterium resistant to being killed by an antibiotic such as ampicillin, neomycin, or tetracycline. Researchers will then grow the transformed bacteria in **growth medium** that contains the appropriate antibiotic, ensuring that only plasmid containing bacteria survive. The transformed bacteria will then grow and divide, producing many duplicate daughter cells that contain the additional genetic information contained in the plasmid.

Once the transformed bacteria start growing, they will make recombinant protein from the gene that was inserted into the plasmid, just as they manufactured the protein that confers antibiotic resistance. Researchers and production engineers can harvest the recombinant protein for a wide range of uses.

MAKING RECOMBINANT PROTEINS

1. Add recombinant plasmids to bacteria.

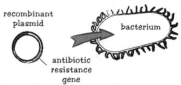

recombinant plasmid

bacterium

antibiotic resistance gene

2. Not all bacteria take up the plasmids.

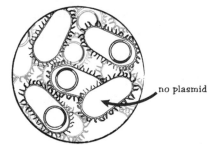

no plasmid

3. Add antibiotic to kill off bacteria lacking antibiotic resistance gene.

4. Surviving culture uniformly composed of bacteria with recombinant plasmids.

plasmids in all bacteria

5. Gene expression leads to desired protein production.

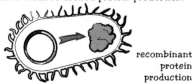

recombinant protein production

GOING FURTHER

Let's examine recombinant protein production by taking a more detailed look at how insulin is now routinely made. In the lab, scientists insert the human insulin gene into a tetracycline resistant expression plasmid and then introduce the plasmid into *E. coli*. The transformed *E. coli* are grown on petri dishes in agar containing the antibiotic tetracycline so that any bacteria without the plasmid die. The bacteria are then transferred to flasks containing nutrient broth that are kept at 37 degrees Celsius and shaken to ensure adequate oxygen circulation. Next the cultures are transferred to fermentation vessels for large scale culture. The fermentation vats are sealed, and the bacteria are grown at carefully controlled temperature with internal stirring. When the bacteria reach the desired density, they are collected and the cell walls are dissolved with enzymes and chemicals. The extract is passed through chromatography columns to purify the recombinant insulin protein. **Column chromatography** is a widely used purification method that separates proteins based on differences in their chemical and physical characteristics, and will be described in more detail in Chapter Ten. At the end of the chromatography step, scientists recover pure human insulin that can be injected into diabetic patients.

Engineering Mammalian Cells

Bacteria are very useful for **molecular cloning**—making many copies of a particular gene—and **expression cloning**—using the information encoded in that gene to produce many copies of a particular protein. Transformed bacteria are essentially DNA and protein factories. However, in order to understand human biology and disease, researchers need to study cells from humans or other mammals. Genetically engineered mammalian cell lines serve many research and production purposes. They can be powerful research models to study the control of gene expression, the normal functions of cells, and the molecular and cellular mechanisms underlying diseases.

Scientists also use genetically engineered mammalian cell lines as protein factories, either on a small scale or a manufacturing level. Mammalian cells offer some advantages over bacteria in human protein expression. Human proteins produced by mammalian cells are often more soluble and structurally sound (folded correctly) than those produced by bacteria cells. These factors are particularly important when it comes to manufacturing human recombinant proteins for use as therapeutic agents.

Scientists grow and maintain mammalian cells in the lab through cell culture, also called **tissue culture**. Scientists can collect cells to culture from a human patient, from a surgery biopsy or just a skin scraping, or from an experimental animal, such as a mouse. The collected sample is treated with an enzyme called trypsin that breaks down the connections between cells in tissue to release individual cells. Researchers culture cells in dishes or flasks in media formulated to mimic the normal physiological environment of the cells, which would include nutrients and growth factors. The cells are cultured at 37 degrees Celsius, which is normal body temperature. Since the same conditions that favor experimental cell growth also favor the growth of contaminating bacteria, yeast, and molds, all work with tissue culture is carried out in sterile laminar flow hoods that blow air away from the work area so as to diminish the risk of contamination. Even under favorable conditions, most cells can only be coaxed to divide a few times before they die. Cancer cells tend to grow well in culture because they have mutated to avoid cell death and many cultured cells are originally from tumors. Once cells have been maintained in culture for a certain length of time—about six months—they are considered a **cell line**.

Cell lines can be frozen in liquid nitrogen and stored until needed. Researchers can thaw out the cells for experiments; although some cells will die during this freezing and thawing process, those that survive will start growing and replicating again once thawed and maintained in favorable conditions. In practice most researchers obtain cell lines that have already been established and characterized, from suppliers such as the American Type Culture Collection (ATCC).

INDUSTRY NOTE

Large companies often have banks of frozen cells stored in different geographic locations as a precaution against losing valuable cell lines. One cell bank might be kept on the east coast, one on the west coast, and a third in another country.

Researchers can insert genes into cultured mammalian cells with plasmids in a manner similar to that in which they transform bacteria. Scientists introduce plasmids into eukaryotic cells by the process of **transfection**, often using lipid reagents that enable DNA to cross the lipid plasma membrane that surrounds the cell. This is a "gentler" method of introducing DNA than that used for bacterial cell transformation, and is necessary because mammalian cells are more fragile than bacterial cells. This method involves the inclusion of the DNA to be transfected into **liposomes**—small, membrane-bounded bodies that are in some ways similar to the structure of a cell and can actually fuse with the cell membrane, releasing the DNA into the cell.

INDUSTRY NOTE

Mammalian cells do not have to be human cells to be suitable research and manufacturing tools. The two most commonly used cell lines for producing recombinant proteins are Chinese hamster ovary (CHO) and mouse myeloma (NSO) cells. Researchers are now moving towards cell culture for preliminary studies of drug testing instead of working with animal models for both ethical and economic considerations. Using only a few mice per month for experiments can add up to as much as $30,000 a year of expense.

Recombinant Proteins in Healthcare

Recombinant proteins have multiple uses as therapeutic agents. Some are recombinant versions of human proteins used to replace or supplement those that are either depleted or lost completely in the body. Examples of these types of supplemental therapies include recombinant insulin for Type 1 diabetes, growth hormone for growth hormone deficiency, and blood clotting factors for hemophilia. Recombinant proteins are also used to modulate immune function in autoimmune disorders. For example, a recombinant version of the signaling protein interferon beta-1 has been an effective treatment for certain multiple sclerosis (MS) patients. Other recombinant proteins stimulate the production of either red or white blood cells, as often required by patients with kidney disease or by those undergoing chemotherapy. In some cases, scientists have found that a viral protein produced using recombinant DNA technology is sufficient to activate an immune response equivalent to that of a traditional vaccine consisting of inactivated whole virus. Yet another category of therapeutically important recombinant proteins is monoclonal antibodies, which have been used successfully to treat a number of diseases including different types of cancer and autoimmune disorders. Monoclonal antibodies will be discussed in greater detail in the next chapter.

PRODUCTS AT WORK: Gardasil

Gardasil, the human papillomavirus (HPV) vaccine manufactured by Merck and approved for preventing cervical cancer, is produced by making a recombinant HPV protein called L1. The recombinant L1 protein molecules reassemble to make virus-like particles (VLP) which mimic real HPV well enough to fool the immune system into making antibodies against HPV.

A growing area in biologic therapeutics is the production of monoclonal antibodies targeting proteins involved in disease. Herceptin is a monoclonal antibody product that was developed to target the 25% of breast cancers that results from the over expression of the HER2 growth factor receptor. When Herceptin binds the HER2 receptor, the growth factor can't bind. Without growth factor binding, growth signals are not sent to the cancer cells, blocking cancer cell proliferation and often triggering their death.

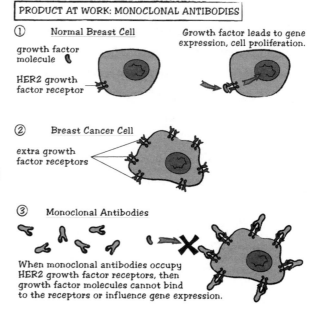

PRODUCT AT WORK: MONOCLONAL ANTIBODIES

① Normal Breast Cell

growth factor molecule

HER2 growth factor receptor

Growth factor leads to gene expression, cell proliferation.

② Breast Cancer Cell

extra growth factor receptors

③ Monoclonal Antibodies

When monoclonal antibodies occupy HER2 growth factor receptors, then growth factor molecules cannot bind to the receptors or influence gene expression.

PRODUCTS AT WORK: Herceptin

To produce Herceptin, scientists engineered a mouse antibody gene so that most of the antibody molecule is human and only a small part, called the variable region, is mouse. This is called a **humanized antibody** and avoids problems with rejection. Most monoclonal antibodies in clinical use today are humanized antibodies.

Pharm Animals

Animals that produce large quantities of milk can be engineered to produce recombinant human proteins only in their milk. This means that the proteins produced in the milk of cows, goats, and sheep are easily renewable. Why is producing human proteins in milk so important? Because the milk secreting system is separate from the rest of its body, an animal can make proteins like insulin or blood clotting factors which would otherwise kill the animal if released into the bloodstream.

Cocktail Fodder

The first transgenic cow to produce a human protein in its milk was Rosie, who produced the human protein alpha-lactalbumin. Rosie's genetically engineered milk was particularly suitable and nutritive for premature babies.

To make a transgenic cow, scientists must first isolate and clone the human gene of interest. They then ligate multiple copies of the gene to a cow DNA promoters. Promoters are stretches of DNA that control the expression of the gene that follows them; gene expression is switched on if certain proteins (transcription factors) bind to the promoter. If those proteins are only produced in cells of a particular tissue, such as mammary gland cells, then the gene will only

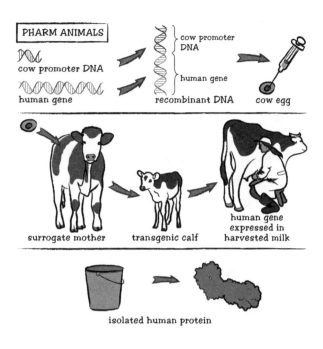

be switched on in that tissue. Scientists then microinject the recombinant DNA into the nucleus of a fertilized cow egg using a fine needle. They next transplant the egg into a surrogate mother who will give birth to a transgenic cow, now expressing the target gene in her milk. The milk can be harvested, and the human protein of interest can be isolated by purification schemes that will be described in Chapter Ten.

PRODUCTS AT WORK: ATryn

The first (and as of January 2012, only) FDA approved biologic drug to be produced using transgenic animals is the recombinant version of a protein called antithrombin, an anticoagulation factor. The product is called ATryn and is produced by GTC Therapeutics.

Disease Models

Transgenic technology, mostly in mice, is crucial for generating models for drug discovery and development. By introducing human disease genes into mice, scientists can create mouse models of that disease to use for developing and testing candidate drugs, and to observe experimentally the roles of genes in development, physiology, and disease.

Cocktail Fodder

Perhaps the most famous transgenic mouse is dubbed the Oncomouse—a mouse containing human oncogenes, or genes involved in turning a normal cell into a cancer cell. The Oncomouse was created by Harvard researchers and subsequently licensed to Dupont, and has been invaluable in helping scientists to understand the cancer process and test new anticancer drugs.

Gene Therapy

Researchers have been making progress in gene therapy, which involves inserting genes into selected cells of a patient to correct genetic disease such as cystic fibrosis. Cystic fibrosis is caused by a defect in the gene which codes for a protein that helps to regulate fluid balance within cells, leading to thick, viscous secretions in cells that line the airway. Treating a cystic fibrosis patient with gene therapy would mean delivering a normal copy of this gene to those patients. One of the major challenges in getting gene therapy to succeed is finding a way of transferring the gene of interest into the target cells. The vector most often used for this transfer is a type of virus knows as adeno-associated virus (AAV).

For many years, clinical researchers have struggled to make gene therapy both safe and effective. Therapies that were effective tended to have unacceptable safety risks, while therapies that were safe proved to be only temporary. Recent advances in vector design have led to the first approval of a gene therapy treatment for a rare genetic condition, lipoprotein lipase deficiency (LPLD). Patients suffering from LPLD are unable to break down fat globules in the bloodstream, resulting in fat-clogged blood vessels in the pancreas and gut, often leading to debilitating pancreatitis attacks. Glybera, approved in July 2012 by the EMA for the treatment of LPLD, is a gene therapy treatment that uses an AAV to deliver the missing lipoprotein lipase gene to LPLD patients. It is the first treatment of its kind to be approved in the U.S. or in the European Union.

Cocktail Fodder

In 1990, 4 year old Ashanti DeSilva received the first successful gene therapy with an adenovirus vector. Ashanti suffered from severe combined immunedeficiency disease (SCID), a genetic disease caused by a defective copy of a gene called adenosine deaminase (ADA). Scientists were able to engineer an adenovirus containing the normal ADA gene to treat her white blood cells, and Ashanti is now a normal, healthy adult.

INDUSTRY NOTE

Genetic engineering has brought us such entrepreneurial endeavors as Genetic Savings & Clone, a commercial pet gene banking and cloning company that delivered the world's first commercially cloned cat, Little Nicky, to a Texas woman for a reported $50,000. Genetic Savings & Clone closed in 2006, though it is still possible to buy a genetically engineered hypoallergenic cat for consumers willing to pay the price. Not yet for sale, but in development in the lab, are transgenic mice with fur that changes color depending on what they eat. It's conceivable to imagine future generations of teenagers being able to change their hair and eye color to suit their mood with the switching on and off of genes instead of with hair dye and contacts.

EXCERPT FROM BIOTECH PRIMER BLOG

Synthetic Life?

The phrase "synthetic life" has been on the lips of many after the publication of a high profile article in the journal *Science* describing the creation of a "synthetic cell". This particular piece of work, carried out by scientists at the J. Craig Venter Institute in Rockville, MD and San Diego, CA, is the culmination of efforts initiated in 1995. Briefly, the researchers used DNA synthesizing technologies to make the genome—the entire DNA content of the organism—of one type of bacterium in the lab, and transferred that genome into a second type of bacterium, whose own genome had been removed. The recipient cell then proceeded to follow the instructions encoded by the replacement genome, essentially converting it to a new type of bacteria.

Why the big fuss? The experiment achieves what scientists call "proof of principle"—it demonstrates that in principle, it is possible to "design" an entirely new genome, or combinations of genes, in the lab, transfer this genome into a bacterial cell, and essentially create a cell with a set of "user defined" functions. Current genetic engineering technology typically involves the transfer of only one or a few genes into a bacteria or other cell type. The ability to transfer an entire genome essentially means the ability to custom design miniature bioreactors for producing fuel or pharmaceuticals, potentially achieving higher levels of productivity.

The Institute next plans to attempt to transfer a synthetic genome into an algae cell—a potentially much more difficult task, since, believe it or not, algae cells are much more complex than bacteria. But because they can perform photosynthesis—convert the sun's energy into sugar—they would make ideal starting points for alternative fuel producing bioreactors.

Excerpt From Biotech Primer Blog Continued From Previous Page

But did the Institute create "synthetic life"? Craig Venter himself denies the claim, clearly stating that he is not "creating life from scratch" since success depends on a pre-existing cell. A better phrase might be "highly processed life". But that doesn't make for quite as catchy a headline.

http://www.BiotechPrimerBlog.com

In this chapter, we have described how scientists genetically engineer cells to produce a particular protein. In the next chapter, we will describe the human immune response, and discuss some of the ways in which biotech companies apply an understanding of this response to new drug development.

CHAPTER SEVEN
Immune Response: Defender Extraordinaire

The immune system is the body's primary line of defense against infection and cancer. Understanding how the immune system functions is crucial to the development of new drugs and therapies for two reasons. First, drugs often have to work in cooperation with the immune system to produce their optimal effects, and many new biotech drugs and biologics, in fact, exploit the mechanics of immune function to work. Second, a great number of diseases are either diseases of the immune system itself, or they arise from a malfunctioning immune system. For these reasons, we'll take a thorough look at the immune system and how knowledge of its mechanisms is being employed by the biotech industry.

> **Understanding how the immune system functions is crucial to the development of new drugs and therapies.**

Immune System Cells

All the cells in the immune system come from a single population of identical cells—**hematopoietic stem cells (HSCs)**. HSCs reside in the bone marrow where they divide to produce identical stem cells, or **progenitor cells**, along with niche cells, particularly **stromal cells** and **osteoblasts**, to help regulate the HSCs and their differentiation. **Hematopoiesis** takes place mostly in the bone marrow and proceeds by an ordered sequence of events via progenitor cell lines, ultimately giving rise to all the different types of blood cells. You may be most familiar with red blood cells (**erythrocytes**), white blood cells (**leukocytes**), and **platelets**, but we will also discuss specific sub-classes of white blood cells, including **B- and T-cell lymphocytes**, **neutrophils**, and **macrophages**, to name a few.

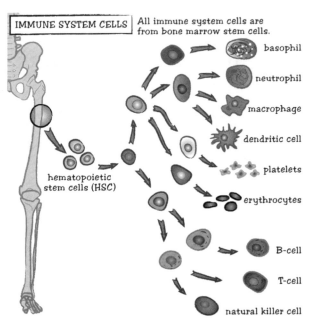

IMMUNE SYSTEM CELLS — All immune system cells are from bone marrow stem cells.

basophil
neutrophil
macrophage
dendritic cell
platelets
erythrocytes
hematopoietic stem cells (HSC)
B-cell
T-cell
natural killer cell

Red blood cells, or erythrocytes, carry many copies of the protein hemoglobin. Hemoglobin binds and transports oxygen throughout the body. It is the reaction between oxygen and the iron-containing heme group in hemoglobin that gives red blood cells their distinctive coloration.

Platelets, or **thrombocytes**, are small cytoplasmic bodies that have no nucleus. They circulate in the blood of mammals and are involved in **hemostasis**, or the clotting of blood. If an organism is low on platelets, excessive bleeding can occur. However, an over abundance of platelets can cause blood clots to form in vessels, causing heart attacks or stroke.

Cocktail Fodder

Most people don't think of the skin as an organ, much less as a part of the immune system. In fact the skin is the largest organ in the body, weighing in at an average of 10 pounds for a typical adult.

Of the white blood cells, lymphocytes are the heavy hitters. B-cell and T-cell lymphocytes recognize and destroy foreign cells and viruses. B-cells act primarily by designing, producing, and secreting antibodies that tag foreign **antigens** for destruction. Destroying foreign or dysfunctional cells—for instance, cells infected by viruses or cancer cells—is the job of the T-cells. The **helper T-cells** send out chemical messages, called cytokines, to recruit other immune system cells to destroy infected cells, and **cytotoxic T-cells** release toxins that destroy foreign cells. B-cell and T-cell functionality will be described in more detail later in this chapter.

Phagocytes, cells capable of ingesting microorganisms, include **macrophages** that engulf and digest foreign objects such as bacteria. **Granulocytes**, such as **basophils** and **eosinophils**, perform a similar task by secreting granules of chemicals that digest foreign objects. **Neutrophils** are both a granulocyte and a phagocyte; they swallow and digest foreign invaders with the help of granules they release. Dendritic cells play a key role in activating T-cells, while natural killer cells seek out and destroy virally infected cells.

Non-Specific Immune Response

The first lines of defense in the human body are the primary barriers presented to invading **pathogens** such as viruses, bacteria, and fungi. These primary barriers include skin, mucous membranes lining the gut and the lungs, and the hairs in the nostrils. Bodily secretions, too, play a role in the first line of defense, which include chemicals in tears, saliva, nasal secretions, sweat, and strong stomach acids that all act to protect the body from infections. Sneezing and coughing is another way in which the body attempts to protect itself from pathogens.

Cocktail Fodder

Mucous is like flypaper for bacteria and viruses! It's wet and sticky and catches little pests that you don't want in your body.

Foreign pathogens that do make it past these surface barriers will often come up against the immune system's nonspecific immune response of innate immunity. This response defends the host from infection by pathogens in a generic manner. An example of a nonspecific immune response is the inflammatory response described in detail on the next page in the Going Further box. Key players in the nonspecific immune response are macrophages that ingest invaders, neutrophils that ingest and destroy pathogens, and eosinophils that secrete toxic substances that also destroy pathogens.

GOING FURTHER

Inflammation occurs as a response to a wound, irritation, or infection by a pathogen. The primary purpose of the inflammatory response is to minimize invasion by foreign matter like bacteria. The response involves recruitment of many types of immune cells to the site of inflammation, including granulocytes and macrophages.

When you cut yourself, bacteria immediately invade the cut, causing the injured cells to release signaling molecules called **cytokines** and **chemokines** that attract immune cells. The released chemicals establish a gradient, most concentrated at the damaged site, and activate immature immune cells— **monocytes**—to leave the blood stream and enter the

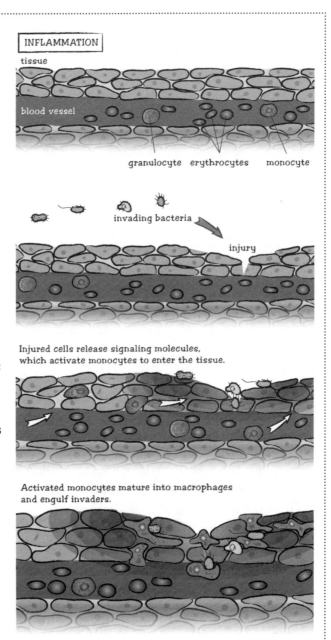

INFLAMMATION

tissue

blood vessel

granulocyte   erythrocytes   monocyte

invading bacteria

injury

Injured cells release signaling molecules, which activate monocytes to enter the tissue.

Activated monocytes mature into macrophages and engulf invaders.

Going Further Continued From Previous Page

tissue. Once in the tissue, the monocytes differentiate into macrophages and eat the bacteria and damaged cells. If the macrophages are able to control the invasion, the cells stop sending "SOS" signals, and the inflammation subsides. The pus that accumulates at the site of an infection is the break down of dead cells.

Classic indications of acute inflammation are redness, swelling, pain, loss of function, and a localized feeling of hotness, caused mainly by increased blood flow to the site of inflammation. Many of the initial responses to acute inflammation are mediated by the release of signaling molecules from white blood cells at the site of inflammation. For instance, **prostaglandins** trigger **vasodilation**, which is a widening of blood vessels that allows increased blood flow. The very short lifetime of these signaling molecules helps to bring the inflammatory response to a quick conclusion.

The return to a noninflammatory condition is known as **resolution** and is an active process, like inflammation. In some cases resolution does not occur, and chronic inflammation may result. This is not associated with the classical signs of acute inflammation and involves destruction or damage to body tissues, often by the macrophages and granulocytes that have infiltrated the tissues to fight the infection. Chronic inflammation can also result from autoimmune reactions that cannot be resolved and may develop into serious autoimmune diseases.

Specific Immune Response

If the nonspecific defenses or surface barriers, such as skin, do not work in fighting off an infection, then the specific immune response, also known as the **adaptive immune response**, kicks in. This is the body's second line of defense to infection and requires use of its "Big Guns", the B- and T-cell lymphocytes.

Lymphocytes are highly specialized cells that continually circulate through the blood and **lymph system** patrolling for foreign invaders or "non-self" cells by spotting antigens that are unknown molecules, usually foreign proteins or chemicals. When you go to the doctor, he feels lymph nodes located in your neck and armpit. If they are swollen, it may indicate that your body is responding to an infection by producing large volumes of white blood cells.

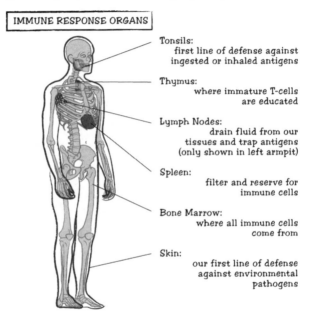

IMMUNE RESPONSE ORGANS

Tonsils:
first line of defense against ingested or inhaled antigens

Thymus:
where immature T-cells are educated

Lymph Nodes:
drain fluid from our tissues and trap antigens (only shown in left armpit)

Spleen:
filter and reserve for immune cells

Bone Marrow:
where all immune cells come from

Skin:
our first line of defense against environmental pathogens

Although the **innate** (nonspecific) and adaptive (specific) immune systems both function to protect against invading organisms, they differ in a number of ways. The adaptive immune system requires some time to react to an invading organism, whereas the innate immune system includes defenses that, for the most part, are always present and ready to be mobilized upon infection. Second, the adaptive immune system is antigen specific and reacts only to the trigger that induced the response. By contrast, the innate system is not antigen specific and reacts equally well to a variety of organisms. Finally, the adaptive immune system demonstrates immunological memory. It "remembers" that it has encountered an invading organism and reacts more rapidly on subsequent exposure to the same organism.

Tricky Terminology

The word "antigen" comes from the words "antibody" and "generator". Examples of antigens include proteins in the cell wall of bacteria, the chemical in the sap of poison ivy, the protective protein coat around viruses, and even proteins associated with human blood types, which is why donor-recipient compatibility is important.

T-Cells

T-cells originate in the bone marrow and mature in the thymus (hence the "T" designation) into either **T-helper cells** or **cytotoxic T-cells** (sometimes called "killer T-cells"). T-cells have unique **T-cell receptors (TCR)** on their surface, designed to recognize antigens. Both T-helper cells and cytotoxic T-cells are activated upon recognizing foreign antigens.

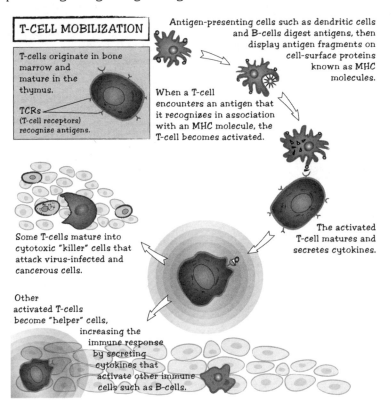

T-CELL MOBILIZATION

T-cells originate in bone marrow and mature in the thymus.

TCRs (T-cell receptors) recognize antigens.

Antigen-presenting cells such as dendritic cells and B-cells digest antigens, then display antigen fragments on cell-surface proteins known as MHC molecules.

When a T-cell encounters an antigen that it recognizes in association with an MHC molecule, the T-cell becomes activated.

The activated T-cell matures and secretes cytokines.

Some T-cells mature into cytotoxic "killer" cells that attack virus-infected and cancerous cells.

Other activated T-cells become "helper" cells, increasing the immune response by secreting cytokines that activate other immune cells such as B-cells.

T-cells can only recognize foreign antigens that are "presented" on the surface of other cells in conjunction with molecules known as major histocompatibility complex (MHC) molecules. An example of this can be seen in virally infected cells. When a virus infects a cell, viral proteins are broken down by cellular enzymes. Fragments of the viral proteins are then displayed by cell surface MHC molecules.

It is these protein fragments that serve as T-cell activating antigens. Other cells of the immune system, including dendritic cells and macrophages, engulf foreign invaders such as viruses, break down their proteins, and display antigen fragments on their surface in a manner similar to virally infected cells. Both types of antigen presentation are required for complete activation of helper and cytotoxic T-cells.

T-helper cells do not have the ability to kill infected host cells or pathogens. They recruit other immune cells to do the job. They accomplish this by secreting signaling proteins called cytokines to activate and direct other immune cells, including cytotoxic T-cells and macrophages. T-helper cells also help to fully activate B-cells, which will then produce antibodies to help clear the pathogen.

Cytotoxic T-cells kill host cells that are infected by viruses or other pathogens, and they kill damaged or dysfunctional cells such as cancerous cells. They are also responsible for the rejection of tissue and organ grafts. **Cytotoxic** T-cells release chemicals called cytotoxins, which create holes and a reaction inside the targeted cells that lead to their death.

Tricky Terminology

Interleukins got their name because they serve as messengers for white blood cells; inter (between)—leukin (white cells). Interferons, though they have multiple functions, were named for their ability to "interfere" with viral replication.

B-Cells

Like T-cells, B-cells originate in the bone marrow, but they also mature there rather than moving to the thymus—hence the "B" designation. B-cells also have specific **B-cell receptors** (**BCR**) on their surface to recognize antigens. B-cells can become activated either by binding to an antigen or by receiving a signal from a T-helper cell, which is T-cell dependent activation.

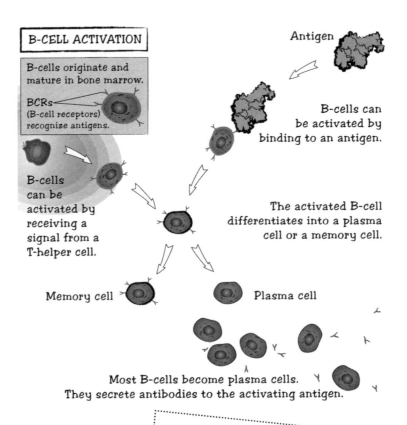

B-CELL ACTIVATION

B-cells originate and mature in bone marrow.

BCRs (B-cell receptors) recognize antigens.

B-cells can be activated by receiving a signal from a T-helper cell.

Antigen

B-cells can be activated by binding to an antigen.

The activated B-cell differentiates into a plasma cell or a memory cell.

Memory cell

Plasma cell

Most B-cells become plasma cells. They secrete antibodies to the activating antigen.

When B-cells are activated, they differentiate into **plasma cells** or **memory cells**. Most B-cells become plasma cells that produce and secrete antibodies. The secreted antibodies target the antigen that initially caused its activation. Some B-cells become long-living memory cells, which circulate and

Cocktail Fodder

You produce an abundance of different B-cells, each able to recognize a different antigen. How is this possible? Each B-cell produces a slightly different receptor. This is accomplished by genetic shuffling of the receptor encoding genes–inside each developing B-cell, segments of the gene recombine with one another, resulting in a unique gene product. Talk about being prepared!

patrol for future invasion of that particular antigen. This function allows the body to respond quickly if infected again with the same pathogen.

Antibodies

Each plasma cell is essentially a factory for producing one type of antibody in response to a specific antigen. Antibodies are "Y" shaped proteins that have a variable region at the top of the "Y" branch that enable it to recognize and bind to a specific portion of an antigen—the **epitope**. There are millions of different antibodies, each recognizing a different epitope. The interaction between an antibody and its epitope is highly specific, and is often likened to a lock and key.

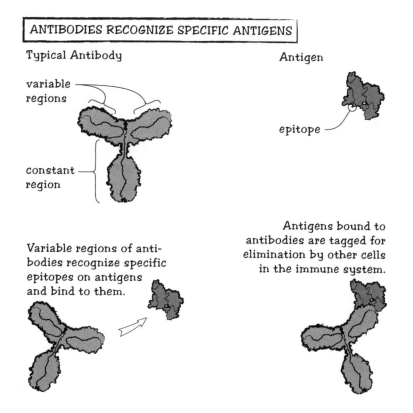

ANTIBODIES RECOGNIZE SPECIFIC ANTIGENS

Typical Antibody

Antigen

variable regions

epitope

constant region

Variable regions of antibodies recognize specific epitopes on antigens and bind to them.

Antigens bound to antibodies are tagged for elimination by other cells in the immune system.

Antibodies attach to foreign antigens and tag them for elimination by other cells of the immune system. The consequences of the secreted antibodies binding to their target antigens may vary. Binding and covering antigens on the cell surface of an invading microbe can cause the microbes to clump together and be incapacitated. Antibodies can signal macrophages to come and eat the invaders. They may also bind to the active site of a toxin to neutralize the toxin.

The activation of the B-cell into plasma and memory cells is called the **primary response**, which lasts for several days or weeks. The concentration of antibody decreases as the plasma cells stops secreting them. Once the infection is eradicated, plasma cells die, but memory cells remain in the body.

If another infection of the same pathogen occurs, then the immune system encounters antigens it has seen before. There is a more rapid response, called the **secondary response**, which is faster because the existing population of memory cells can produce many plasma cells and the appropriate antibody. These destroy the pathogen before it has the chance to cause any symptoms to occur.

Cocktail Fodder

An activated B-cell can produce up to 50,000 antibody molecules each minute, and a macrophage can consume as many as a half dozen bacteria at a time! The word macrophage literally means "big eater". It's no wonder we feel weary when we're sick. Our bodies are actually working very hard when battling infection.

INDUSTRY NOTE

Autoimmune disorders can sometimes be managed by treating patients with biologic drugs designed to either interfere with activating cytokines or to supplement inhibitory cytokines.

Biotech Applications of Antibodies

The ability of antibodies to recognize and bind to one specific antigen has been exploited to great advantage in the biotech industry. Antibodies have been used to great success in both therapeutic and research applications. We will discuss some of these applications later in this chapter. First, let's look at how antibodies to a specific antigen are obtained by scientists.

Cocktail Fodder

Scientists estimate that humans generate about ten billion different antibodies, each capable of binding a distinct epitope of an antigen.

Scientists make **antiserum** by injecting an animal—usually a rabbit, sheep, or goat—with the protein they want to make antibodies against. The rabbit B-cells recognize the protein as foreign and produce antibodies that bind to multiple epitopes on the protein. Scientists will repeat the procedure, and each successive immunization produces increasingly more antibody due to memory B-cells made by the animal. The antibodies generated in this way are called **polyclonal**.

Polyclonal antibodies (or antisera) are antibodies that are derived from different B-cell lines. They are a mixture of antibodies secreted against a specific antigen, each recognizing a different epitope. Because of this, polyclonal antisera often bind to, or cross react with, non-related proteins. This specificity problem was overcome by the development of **monoclonal antibody** technology in the mid 1970's by Niels Jerne, Georges Köhler, and César Milstein, who were recognized for their work with a Nobel Prize in Physiology or Medicine in 1984.

Monoclonal antibody production relies on the clonal selection of B-cells; researchers isolate a single B-cell clone to produce identical antibodies that all bind to the same epitope. Monoclonal antibodies are most often made from mouse B-cells. First, researchers immunize a mouse with the protein antigen. Once the mouse has had time to generate an immune response against the antigen, scientists remove its spleen as a source of B-cells. Since B-cells cannot be maintained in cell culture and would soon die, scientists have to "immortalize" them. They accomplish immortalization by fusing the B-cells to **myeloma** cancer cells. The two types of cells are physically fused together to produce a hybrid cell, appropriately called a **hybridoma**.

The hybridoma cells retain both the antibody-producing capability of B-cells and the cell culture immortality of the myeloma cells. Once researchers isolate hybridoma cells, each one needs to be screened in order to determine which is producing antibodies with the desired characteristics.

INDUSTRY NOTE

The specificity of monoclonal antibodies in comparison to polyclonal antibodies makes them particularly useful for therapeutic applications. In fact, all antibody-based therapies rely on monoclonal antibodies, whereas for many research purposes, polyclonal antibodies will suffice.

Humanizing Mouse Antibodies

Though monoclonal antibody (mAb) technology was invented in the mid 1970s, its promise in human therapy was not realized for almost 20 years. This is because the human immune system identified the mouse antibodies as foreign, and the resulting **human anti-mouse antibody (HAMA)** response decreased plasma levels of the monoclonal antibodies to ineffective levels, and posed a threat of serious allergic reactions. The success of monoclonals in immunotherapy today is due to the ability to reduce HAMA responses by making mouse antibodies more closely resemble a human protein.

After isolating a hybridoma clone expressing a monoclonal antibody of therapeutic interest, scientists can manipulate the gene encoding that antibody. First, scientists isolate the small portion of the mouse antibody gene that encodes the region of the antibody required for antigen recognition. This small DNA segment is then spliced into a human antibody gene, using the techniques described in Chapter Six. This hybrid gene is then introduced into mammalian cells for expression. The **humanized** antibodies produced from these cells are not recognized as foreign by the human immune system, and so do not engender a HAMA response. It is now also possible to generate fully human monoclonal antibodies from transgenic mice engineered to express the human antibody gene.

INDUSTRY NOTE

Nearly all FDA approved immunotherapeutics are humanized. Chimeric antibodies generally have more of the mouse genes (~20-25%) than the humanized antibodies (~5-10%). Humanized antibody molecules elicit a greatly reduced HAMA response.

Researchers have created monoclonal antibodies that have been approved to treat a number of diseases, including cancer, cardiovascular disease, inflammatory diseases, macular degeneration, transplant rejection, multiple sclerosis, and viral infection. This type of treatment is called monoclonal antibody therapy.

PRODUCTS AT WORK: Rituxan

In an earlier section we mentioned the therapeutic Rituxan, which was the first antibody therapeutic to be approved and is used in the treatment of the blood cancer non-Hodgkin's lymphoma (NHL). Now we'll take a closer look at the disease and the therapeutic action of Rituxan.

NHL is the fifth most common cancer in the U.S.. It is incurable but can be managed by immunotherapy and chemotherapy. In 80-90% of NHL patients, the cancer affects B-cells that express the transmembrane glycoprotein CD20. The function of CD20 is still unknown, but there is evidence to suggest that it plays a role in controlling calcium ion levels inside the B-cells.

Rituxan is a humanized antibody that targets CD20. It binds specifically to CD20 on the surface of both normal and tumor B-cells. Because it does not bind to other surface molecules, there is a reduction of potential side effects. Rituxan targets normal and tumor B-cells for destruction in two ways. First, Rituxin binding seems to trigger a signal that leads to **apoptosis**, a form of cell suicide. Second, once bound cells of the patient's own immune system recognize the Rituxin-B-cell complex as foreign and attack and destroy it.

Although normal B-cells are also killed, they can be replaced quickly since these bone marrow cells do not express CD20. Antibody secreting plasma cells also lack CD20 and so are not affected either. The ability of Rituxan to clear CD20 expressing B-cells in multiple ways probably accounts for its effectiveness in treating NHL and also its versatility in treating autoimmune diseases such as rheumatoid arthritis.

HIV Vaccine: Steps Closer to an Elusive Goal

Developing a vaccine to thwart human immunodeficiency virus (HIV), the virus that causes AIDS, has been a major goal of researchers since the virus was first identified in 1983. There have been many false starts and a few signs of hope, but no real success stories. The latest development in this arena takes an entirely new approach to the problem.

Traditional vaccines rely on injecting a weakened or inactivated version of the virus into the patient, in hopes of training the patient's immune system to recognize the virus and consequently to be prepared to defend against it in the event of an actual infection. One of the key ways in which the immune system can defend against a virus is by producing proteins called antibodies that recognize portions of the virus particle, bind those recognized portions, and prevent the virus from entering its target cell—rendering the virus harmless, or "neutralizing" it. This approach has not been successful as an HIV immunization strategy, in part because of the high rate of mutation that the virus exhibits—in effect, the immune system is trained to recognize Version One of the virus, and the patient becomes infected with Version Two.

In recent years, various investigators have identified what are referred to as "broadly neutralizing antibodies"—antibodies that are capable of neutralizing a broad range of different HIV strains. These antibodies arise by chance in a minority of HIV-infected individuals, and are thought to be able to recognize multiple strains of the virus because they target regions of the virus that mutate at a much lower frequency than other regions of the virus.

Although these antibodies are highly effective against the virus, they only arise in a small number of patients—and only after the infection has already taken root. What if scientists could isolate the genes that provide the instructions for making these antibodies, and deliver them directly to

Excerpt From Biotech Primer Blog Continued From Previous Page
patients—affording the chance to produce their own powerful antibodies capable of defending against multiple strains of HIV, prior to infection?

David Baltimore's lab at the California Institute of Technology has taken the first steps towards doing just that. Using gene therapy technology, the team introduced the genes that code for these broadly-neutralizing antibodies into muscle cells of mice that have been genetically altered to contain a human immune system—and thus susceptible to HIV. These mice muscle cells then proceeded to make large quantities of the powerful antibodies—rendering the mice resistant to HIV infection. The lab refers to this technique as "vectored immunoprohylaxis (VIP)"—and in the words of Dr. Baltimore, "The advantage of VIP is simply that it works." After almost thirty years of trying, this is indeed quite an accomplishment. Human clinical trials are expected to begin in 2012.

http://www.BiotechPrimerBlog.com

In this chapter, we have examined the inner workings of the human immune system, and learned how biotech companies apply this knowledge towards developing new products. The next chapter will provide a detailed overview of the drug discovery process.

CHAPTER EIGHT
The Science Of Discovery

A Brief History

For thousands of years, people have sought natural sources to alleviate various human ailments. As explained in Chapter One, aspirin was first extracted from the bark of the willow tree, and was described by Hippocrates in the 3rd century BC for treating aches, pains, and fever. Native Americans, too, used willow tree bark for centuries. Digitalis, or digoxin, is extracted from the foxglove plant and is used to treat cardiac arrhythmias and other heart conditions. Native Peruvians treated malaria with bark from the cinchona tree, which grows in the Andes. In the 19th century, scientists extracted quinine from cinchona bark and identified it as an anti-malarial drug.

By the mid 20th century, scientists were developing drugs in the lab using trial-and-error methods. For instance, to find an alternative to aspirin for treating pain and inflammation, British scientists at the Boot Laboratories tested several hundred unrelated chemicals in guinea pigs to look for drugs that would reduce inflammation. This approach to drug discovery was time consuming and resource intensive, and not always successful. The type of companies that tended to succeed

Cocktail Fodder

Researchers test a drug's efficacy in relieving pain by a method called "the rat tail test". They secure a rat, placing its tail near a heat source. If the rat does not try to raise its tail away from the heat as the heat increases, the pain medication probably works! If the rat raises its tail to escape discomfort, researchers know to stop the test and return to the drawing board.

using this method were large, publicly held companies, giving rise to the term "Big Pharma".

Beginning in the 1980's, scientists began to approach drug discovery a little bit differently, adopting something known as **rational drug discovery** or **mechanism-based drug design**. In this strategy, researchers first sought to understand the disease mechanism at the cellular level, and, if possible, identify the cellular mechanism associated with the disease. Many diseases can be traced either to the underproduction of a particular protein (e.g., insulin in Type I diabetes); the overproduction of another type of protein (e.g., the growth factor receptor HER2 in certain types of breast cancer); or the production of a mutated version of a protein that no longer functions properly (e.g., a mutated version of a tumor suppressor that no longer controls cell division, leading to cancer.) By understanding what protein or proteins are associated with a particular disease, researchers can then design drugs to specifically target or replace those proteins. Although still labor and resource intensive, this strategy of drug development tends to be better at producing highly effective, specific drug therapies for a given indication than the old trial-and-error approach.

INDUSTRY NOTE

Further refinement of the rational drug design approach introduces structural information about the target protein. This enhancement led to the invention of HIV protease inhibitors. The HIV protease is a viral protein whose function—to cleave other HIV proteins prior to viral assembly inside of a host cell—is critical to HIV's survival and ability to infect other host cells. Once this was understood, researchers realized that by inhibiting the protease's activity, they could stop the spread of the virus. Highly effective inhibitors were designed in part through an understanding of the protease's structure.

In the 1990's, chemists developed huge collections—libraries—of related chemical compounds using **combinatorial chemistry**. Combinatorial chemistry is an approach that enables researchers to develop new molecules by synthesizing every possible chemical derivative of an existing molecule. This technique can often lead to chemical libraries containing hundreds of thousands of unique molecules. A related approach termed directed evolution is used to generate new biologics—specific mutations are introduced to a gene sequence in order to generate new, but structurally related, proteins.

Tricky Terminology

An **assay** is the scientific term for test. In drug discovery, assays are tests developed by scientists to measure the potential efficacy of a drug candidate.

These chemical compound and biologic libraries are created for use in **high-throughput screening** to find the best drug for a particular disease. For example, researchers might screen a large chemical library to see which molecules best inhibited the HIV protease. In some sense, we can think of this as an integrated approach to drug discovery—one that combines the previously used trial-and-error approach of screening a large number of compounds with the newer mechanism-based approach of designing those screening assays based upon an understanding of the disease mechanism.

Cocktail Fodder

When Pfizer scientists tried to develop a new drug to treat high blood pressure, they took the drug sildenafil into clinical trials. The drug failed to control blood pressure, but many of the male trial volunteers noted a dramatic side effect of the drug. Pfizer renamed their discovery Viagra and marketed the first drug for erectile dysfunction.

Serendipity has always played a role in drug discovery. When Alexander Fleming went on vacation and returned to find his petri dishes of bacteria contaminated with mold, he didn't throw them out. Instead he noticed that mold seemed to prevent bacterial growth, and he surmised that the *Penicillium* mold must be producing something that kills bacteria. Quite by accident, Fleming had discovered penicillin, the first antibiotic.

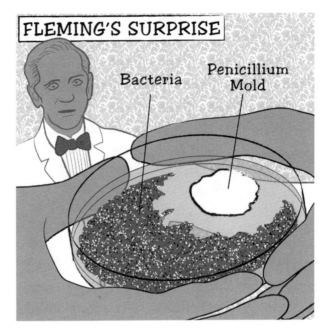

Identifying the Target

The first and arguably most important step in a successful drug discovery program is working out the molecular basis of the disease—in other words, determining the **drug target**. Target identification relies on many years of basic research devoted to understanding the biology of the disease.

Let's look at coronary artery disease (CAD) as an example. The most serious and life threatening cardiovascular diseases are caused by atherosclerosis, which is the build-up of a cholesterol plaque inside an artery that restricts blood flow and can lead to heart attack and stroke. Atherosclerosis can be caused by deposits of a particular type of cholesterol known as low-density lipoprotein, commonly known as LDL or "bad cholesterol".

Although the typical American diet is unfortunately high in cholesterol, this is not the only source of cholesterol for our bodies. Cholesterol is also synthesized in the liver, and while many individuals can control their cholesterol levels through diet and exercise, others have overactive cholesterol synthesis enzymes and will be prone to high cholesterol levels regardless of diet. By identifying the key enzyme required for cholesterol synthesis—HMG-CoA reductase—scientists were able to successfully developed drugs to block the synthesis of cholesterol in at-risk patients.

Cocktail Fodder

The drugs that inhibit HMG-CoA reductase are collectively called statins, which include Crestor, Zocor, and Lipitor.

Types of Targets

Noted researcher Sir James Black, with Gertrude Elion and George Hitchings, who won the 1988 Nobel Prize in Physiology or Medicine for discovering the beta blocker propranolol to treat hypertension and heart disease, once commented, "The most fruitful basis for the discovery of a new drug is to start with an old drug." Black had learned through his extensive research that some classes of drug targets are more likely to be successful than others; if a drug against a particular target works well, it is likely that a similar drug against a similar target will also work well.

Among the various target classes, G protein coupled receptors (GPCRs) have been the most successfully targeted class in modern therapeutics. Recall from our earlier discussion of cell signaling (Chapter Two) that the G protein coupled receptors are widely used in almost all cells to regulate a variety of processes, from blood pressure to nerve transmission to stomach acid secretion. Receptor proteins and ion channels have also been successfully exploited as drug targets. In fact, more than 50% of all approved drugs target either GPCRs, receptors, or ion channels. Finally, although there are currently relatively few drugs approved that target the active site of protein kinases, kinases are attracting huge attention from biotech and pharmaceutical companies as potential drug targets for cancer and many other diseases. As discussed in Chapter Two, protein kinases play a critical role in growth factor signaling cascades.

PRODUCTS AT WORK: GPCR Targeting Drugs

GPCR—targeting drugs include beta blockers (epinephrine beta receptors) such as Atenolol, H2 blockers such as Zantac and Pepcid, and H1 blockers such as Claritin and Allegra (histamine receptors), Imitrex for migraine (serotonin receptors), and Zyprexa for schizophrenia and bipolar disorder manic episodes (dopamine receptors).

Validating the Target

Once a potential drug target has been identified, researchers will try to validate the target by determining whether the target plays a key role in the disease process and whether targeting it is likely to be both safe and effective. Target validation is a very important step in the drug discovery process, since research and development gets progressively more expensive. "Fail early and fail often" is an industry motto. Why? If a drug is ultimately going to prove unsuccessful, millions of dollars can be saved if this failure is realized early on.

Target validation will most often include cell-based assays (*in vitro* testing) and animal models (*in vivo* testing). Since the goal of many therapeutic interventions is to inhibit the activity of the selected target, many validation assays attempt to measure the effects of inhibition. In some cases, a selected target may play a role in disease progression—but if it is inhibited, another cellular protein will simply take its place, nullifying the potential therapeutic effect of an inhibitor. In other cases, inhibiting a selected target may have the desired therapeutic effect—halting cancer cell growth, for example—but may also result in unexpected side effects, such as the death of healthy cells resulting in a heart attack.

TARGET VALIDATION: RNAi

In vitro: cell models

1. Does the target play a key role in the disease process?

2. Is targeting it likely to be effective and safe?

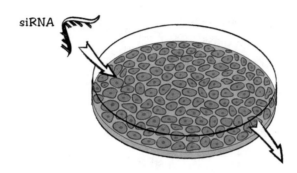

siRNA

- Will cancer cells die?

- Will nerve cells stay alive?

- Will beta cells make more insulin?

- Will liver cells make less cholesterol?

One of the most popular ways of testing the effects of inhibition in cell based assays is through the use of RNAi, described in detail in Chapter Four. RNAi is an effective way to quickly determine the results of blocking the production of a particular protein, thus mimicking the effects of a strong inhibitor.

If the cell models show promise, the researchers will move on to animal models, most likely designing experiments using so-called "knockout" mice— mice in which a particular gene has been disrupted. Researchers can ask similar questions to those asked in the cell model, but on the scale of the whole animal: do the experimental mice still get cancer, Parkinson's disease, diabetes, or heart disease when the target gene is silenced or absent? The animal model also provides valuable information about targeting safety that might not be addressed in cell models, because it is possible to examine the effects of gene targeting on the whole organism.

Tricky Terminology

The terms *in vitro*, *ex vivo*, and *in vivo* are ways to designate where an experiment is taking place. All terms derive from Latin roots. *In vitro* translates to "within glass" and refers to studies carried out in cells grown for extended periods of time in the lab. *Ex vivo* translates to "out of the living" and refers to experiments carried out on cells, tissues, or organs that have recently (typically defined as within 24 hours) been removed from a living organism. *In vivo* translates to "within the living" and refers to experiments carried out on living animals. A modern (1989) term, *in silico*, is sometimes used to refer to computer simulations or computer modeling experiments.

Theraputic Choices: Small Molecule Versus Large Molecule Drugs

The next step in drug discovery involves designing a targeting strategy. This usually comes down to a choice between a small molecule drug and a large molecule drug also known as a biologic drug. Each approach has its particular advantages and disadvantages.

Small molecule drugs can usually cross cell membranes and enter cells, allowing them to target proteins inside cells. Small molecules are often able to cross the blood-brain barrier. Their specificity to their target can vary—it may be good, or even very good, but in some cases can be only fair, potentially resulting in unanticipated off-target toxicity. Small molecules can be taken orally, but once inside the body they generally have a short **half-life**, meaning they are quickly broken down by drug metabolizing enzymes in the liver.

Large molecule drugs, in contrast, usually cannot cross cell membranes or the blood-brain barrier. Biologics target the surface of cells or pathogens. Biologics generally have extremely high specificity, meaning that they are less likely to interfere with proteins other than the target protein, resulting in a lower likelihood of off-target toxicity. Since biologic drugs are proteins, this means they cannot be taken orally—digestive enzymes would break them down. Once inside the body, however, they are quite stable relative to small molecule drugs because they are not broken down by the liver.

Tricky Terminology

The **half-life** of a drug is a measure of how long it remains present in the bloodstream or in its target tissues. The name derives from the fact that it is a measurement of the amount of time that it takes for the drug concentration to be reduced to one half of its initial levels.

INDUSTRY NOTE

Small molecule drugs are synthesized by chemists in the laboratory, whereas large molecule drugs are produced by cells. The use of cells to produce biologic drugs—biomanufacturing—adds considerable time and expense to the production process. Biomanufacturing will be discussed in more detail in Chapter Ten.

PRODUCTS AT WORK: Therapeutic Options

Some diseases can be targeted by both small molecules and biologics. This is particularly true for growth factor receptors, since part of the receptor is outside of the cell (extracellular) and part of it is inside of the cell (intracellular). Physicians can target the HER2 growth factor receptor with a large molecule monoclonal antibody drug such as Herceptin, as well as with small molecule inhibitors such as Tykerb. Similarly, the epidermal growth factor (EGFR) receptors can be targeted by anti-EGFR antibodies such as Vectibix, as well as small molecule inhibitors such as Iressa.

Assay Development

To screen potential drugs against a given target, scientists design assays to identify the drug candidates with the most potential. The assay must be fast but accurate as well as amenable to scale-up, so that thousands, or even millions, of compounds can be screened efficiently. Often researchers develop assays that produce a fluorescent signal or color change because fluorescence is easily measured, relatively inexpensive, and safe, and works with many different assay designs.

Scientists frequently use small molecule inhibitors to target cellular enzymes. Enzymes have a pocket called an **active site**, a particular location where chemical reactions are catalyzed. Active sites have

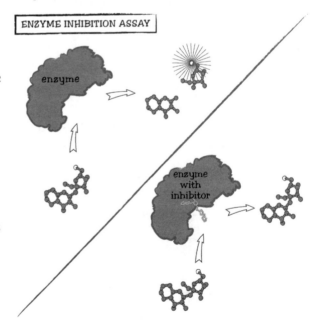

ENZYME INHIBITION ASSAY

enzyme

enzyme with inhibitor

specific structures designed to bind only the enzyme's **substrate**, the molecule it interacts with. Researchers can chemically modify substrates so that they give off a fluorescent signal if the enzyme acts on them. Enzymes treated with affective inhibitors will not successfully catalyze the reaction, and thus the substrate will not emit fluorescence or will emit it at reduced levels. This provides a quick assay readout.

If the target is a receptor, researchers can employ a similar strategy. Scientists genetically engineer cells so that they produce a fluorescent signal when a particular receptor is activated. For example, a cell could be engineered to produce a green fluorescent protein in response to activation of the breast cancer associated HER2 receptor. Researchers would then add a drug candidate to inhibit HER2, followed by an appropriate HER2-activating growth factor, and measure fluorescence levels. If fluorescence is seen the candidate is not blocking the HER2 receptor and is proven to be an ineffective inhibitor. However, if no florescence is measured the candidate may be an effective inhibitor and researchers will continue testing this candidate.

GOING FURTHER

To select the best possible enzyme or receptor inhibitor, researchers need to screen as many potential candidate compounds as possible. In the case of small molecule drug candidates, the best way to screen as many compounds as possible is often through the development of a **combinatorial library**.

A combinatorial library is assembled like a pyramid. First, a small but diverse set of ~20 chemical compounds is attached to small beads. These are the "scaffold" molecules on which the library will be based. Each set of beads is split into ~50 batches and reacted with a different chemical reagent. This generates ~1000 first round compounds attached to beads. Next, these beads are pooled together and split into ~50 batches, then reacted with a different set of ~50 second round reagents. This generates a library of ~50,000 second round compounds. Researchers can repeat the process until millions of different compounds are generated.

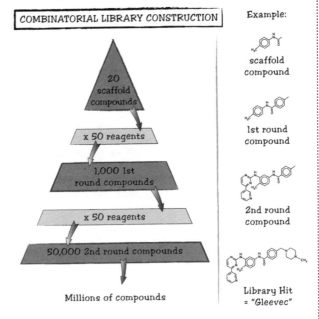

COMBINATORIAL LIBRARY CONSTRUCTION

20 scaffold compounds

x 50 reagents

1,000 1st round compounds

x 50 reagents

50,000 2nd round compounds

Millions of compounds

Example:

scaffold compound

1st round compound

2nd round compound

Library Hit = "Gleevec"

High Throughput Screening

With so many possible chemical compounds to test as drug candidates for a particular target, scientists must maximize efficiency of screening. When developing assays, they aim for high throughput screening. High throughput screening assays usually take place in wells of a microtiter plate containing 96, 384 or 1536 wells. Because of scientists' ability to generate extremely large numbers of small molecule drug candidates through the use of combinatorial chemistry, high throughput screening is a technique most often associated with small molecule drug discovery.

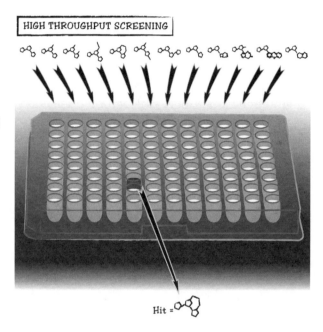

HIGH THROUGHPUT SCREENING

Hit =

To screen for an inhibitor of enzyme X, researchers add the enzyme to each well in a buffer that allows the enzyme to work properly. Next researchers add a different chemical compound to each well, one that has some similarity to the enzyme's natural substrate. The idea is that the inhibitor is expected to fit into the enzyme's active site, preventing the natural substrate from binding. After the test compounds have been added, the fluorescently labeled substrate is added to each well. All of those wells in which the enzyme is active will fluoresce, but any wells in which the enzyme is inhibited will not fluoresce. Any "blank" wells represent potential inhibitors. These potentially promising compounds are referred to as "hits" and merit further investigation.

PRODUCTS AT WORK: Gleevec

The poster child of rationally designed anticancer drugs is Gleevec, used to treat chronic myelogenous leukemia (CML). A rearrangement of two chromosomes in lymphocytes causes CML. This translocation causes two genes—bcr and abl—to join together to form a new hybrid gene. The hybrid gene produces a chimeric protein—part Bcr protein and part Abl protein. The Abl protein is a protein kinase involved in controlling cell division and survival. In normal lymphocytes, Abl is usually in the "off" state, waiting for the appropriate stimulus to switch it on. When lymphocytes are exposed to growth factors, Abl is switched on briefly and then switched off again. While switched on Abl adds a phosphate, in a process known as phosphorylation, to target proteins which is a signal that tells the cell to multiply.

Abl's attachment to Bcr in CML causes it to be permanently switched on, regardless of whether growth factors are present or not. The permanent "proliferate and survive" signal leads to uncontrolled lymphocyte growth and, eventually, leukemia, especially CML.

Researchers screened a library of compounds looking for something that would fit into the active site of Bcr-Abl, potentially blocking its ability to remain switched on. They found one called imatinib, which blocked Bcr-Abl's kinase activity and stopped the proliferation of cancer cells in the lab and the growth of tumors in animals. By examining the structure of a Bcr-Abl crystal with x-rays, researchers observed that imatinib fit into the active site of the Abl part of the molecule. The precise fit prevented imatinib from blocking most other kinases, resulting in fewer side effects.

The success of treating CML with imatinib—trade name Gleevec—was unprecedented, with more than 90% of patients remaining leukemia free five years after trials started. In fact, because Gleevec was so successful in clinical trials, patients taking placebo or another drug were switched to Gleevec before the end of the trials.

Products At Work Continued From Previous Page

Although Gleevec blocks both Bcr-Abl and normal Abl, it has very little impact on normal lymphocyte function, providing evidence that tumor cells are far more dependent on their mutated oncogenes than non-tumor cells are on their corresponding normal genes. Following quickly on the success of Gleevec, researchers identified seven other kinase blocking drugs that have been approved by the FDA for cancer treatment. Most of these target the receptor kinases that are overexpressed or mutated in cancer.

INDUSTRY NOTE

More than one hundred other kinase targeted cancer drugs are in clinical trials, and many of these target the intracellular kinases that are mutated in cancer. Researchers consider kinases to be "druggable" because they understand a lot about their structure and can synthesize molecules to fit into their active sites. Other protein molecules, such as the transcription factors downstream of the kinase cascades, play important roles in cancer but are often considered "undruggable" because scientists don't yet know enough about their structure or how to block their function.

From Hit to Lead

At the end of the high throughput screening process, researchers will ideally have identified a number of compounds that show promise as drug candidates. These promising candidates are called "hits". In order to become "leads" worth moving forward into animal testing, hits must be subjected to rigorous *in vitro* testing in order to answer the three following questions:

> ### Cocktail Fodder
> In the 1980's, great strides in drug discovery and development coincided with quickly rising rates of HIV infection and diagnoses of the disease AIDS. Treatments for AIDS, starting with AZT in 1987, were among the mostly quickly developed drugs ever in the industry.

1. Is the candidate safe?
2. Is the candidate specific to its target?
3. Is the candidate effective in treating the disease?

If the candidate is designed to inhibit an enzyme—a protein kinase, for example—it should only inhibit that particular target kinase and not interfere with any other kinases since many protein kinases play essential roles in normal cell growth. If the compound is not adequately specific, then the drug might have unwanted side effects, which may be significant enough to make the drug unsafe. Protein kinases, like other families of enzymes, share common sequence elements and structures; therefore, some degree of cross-reactivity is expected. However, researchers aim to select a drug that inhibits the target enzyme at a much lower concentration than would affect other related enzymes. This provides them with a **therapeutic window**, a concentration range within which the drug is both effective and safe. Researchers often perform specificity studies by testing the drug for inhibition with a large panel of related enzymes.

INDUSTRY NOTE

Research support companies supply panels of protein kinases for bulk specificity testing. In fact, some of these companies also offer the service of testing drugs inhouse, then sending the results to the researcher.

The next consideration for a compound in early testing phases is whether it is likely to be effective against the disease. Scientists will test a drug's **efficacy** in cells, starting with cells from human patients or animals designed to mimic critical features of the disease, before they move on to studies in whole animal models. From cell model efficacy experiments, researchers can derive an **EC50 value**, which is the drug concentration at which half of the maximum desired effect is observed. For instance, if the desired effect of the drug is to kill tumor cells, then researchers know its EC50 value when they observe that half of the tumor cells have died. Cancer cell models are particularly well suited to these studies because of their ability to grow in tissue culture.

Scientists can also begin to address the question of whether a drug is likely to be safe in cell based models, although the majority of safety testing comes in animal studies. Many drugs bind to, and interfere with, the function of a potassium ion channel expressed in the heart called hERG. This interference causes a change in heart muscle function, which can lead to life threatening arrhythmias. A number of drugs have been withdrawn from the

Cocktail Fodder

The acronym hERG stands for "human Ether-a-go-go Related Gene". (Really!) The "ether-a-go-go" gene was first identified in fruit flies, and scientists observed that mutations in this particular gene caused fruit flies to shake their legs like go-go dancers. Needless to say, they named the gene in the 1960's when go-go dancing was in its heyday.

market because of hERG blocking. To test drugs for hERG blocking, researchers use cell lines engineered to express hERG. The experimental drug is then added to the cells, and instrumentation capable of measuring electric current is used to determine whether or not the hERG ion channel is being inappropriately activated.

Biomarkers

An important tool that can be used throughout drug discovery and development are **biomarkers**. A biomarker can be defined as a specific, measurable physical trait used to determine or indicate the effects or progress of a disease or condition. A classic example of a biomarker is LDL cholesterol, which has been linked to heart disease in numerous studies. Cholesterol is an ideal biomarker because 1) it is found circulating in the blood so it is very easy to obtain from the patient; 2) it is an early indicator of the disease—a patient can have elevated LDL decades before developing significant atherosclerosis; and 3) it changes with disease progression. This change in LDL can be used as a clinical endpoint in lieu of waiting potentially decades to demonstrate a change in disease progression. This use of biomarkers is very important in running shorter, less expensive clinical trials. Statins, such as Lipitor, which have been commercially available since 1987, are now known to increase life span because patients have been taking them for over 25 years. But the initial clinical trials simply showed a decrease in blood cholesterol.

Obscure Syndrome: Source of Cancer, Diabetes Cure?

Laron syndrome is a disorder that results from an individual's inability to respond to human growth hormone. The lack of response is linked to a mutation in the growth hormone receptor protein, or cell surface protein that receives and transmits chemical signals to cells.

Laron syndrome is not common. However, for the past 22 years, scientists have been studying a small population in the Ecuadorian Andes among which the syndrome is common. The observable features of the disorder include a short stature and prominent forehead. What is remarkable, however, is the observation that Laron's syndrome patients almost never get cancer or diabetes. Even obese patients remained sensitive to insulin and did not get Type II diabetes, which is associated with obesity in normal populations. And this protective effect towards cancer and diabetes was also observed in mice unable to make the growth hormone receptor.

All of this points to a potential new application of the drug Somavert, currently marketed by Pfizer as a treatment for acromegaly, a disorder caused by the production of too much growth hormone, resulting in soft tissue swelling and gigantism. Somavert blocks the action of the growth hormone receptor and has been on the market since 2003. Of course, using the drug as a treatment for a new indication such as certain types of cancer or diabetes would require a whole new round of clinical trials—and the benefits may not outweigh the risks. Like any drug, Somavert has side effects, including potential cardiovascular problems. But the connection is still an intriguing one, and may lead to the discovery and development of entirely new treatments.

http://www.BiotechPrimerBlog.com

CHAPTER EIGHT: THE SCIENCE OF DISCOVERY

In this chapter, we have outlined the critical steps required for successful drug discovery. In the next, we will explore the process by which new drugs gain marketing approval.

CHAPTER NINE
Lab To Marketplace

The Drug Development Process

A successful drug development project starts with the identification of potential new drugs in the lab; proceeds to testing those compounds in animals and then humans, and ends with FDA approval for marketing. During the drug discovery phase, researchers may screen thousands, or even millions, of potential compounds, selecting only a handful for further development; a single one of these drugs might make it to market.

> **Drugs must prove safe and effective in human patients prior to market approval.**

Regulatory Agencies

Drug candidates must overcome a number of hurdles before getting final approval, and all these hurdles concern the drug's safety and efficacy in human patients. The path to drug approval is regulated and overseen by governmental agencies. In the U.S., Congress has passed laws regulating drug and biologic approval, including the Federal Food, Drug and Cosmetic (FD&C) Act in 1938. The intent of this legislation was to ensure drug safety in the aftermath of the death of more than a hundred patients from a poorly manufactured sulfanilamide drug. Congress later amended the Act to include guidelines pertaining to drug effectiveness. It is the most far-reaching piece of legislation covering regulation of drug approval.

Most countries have a regulatory equivalent of the FDA. In the European Union, the European Medicines Agency (EMA) acts as a collective regulatory body for the EU member states so that drug makers can file for drug approval centrally through the EMA. This centralization also assures that the drug approval process is not prejudiced by domestic interests within individual member states.

Preclinical Development

The discovery phase of drug development can last between two years to more than a decade—sometimes longer. How long it takes for a drug candidate to become eligible for testing in human patients depends on many factors, including the complexity of the disease, the chemical characteristics of the drug candidate—simple properties like drug solubility in water can affect the ability to develop them—and the amount of resources in both dollars and manpower that the company invests.

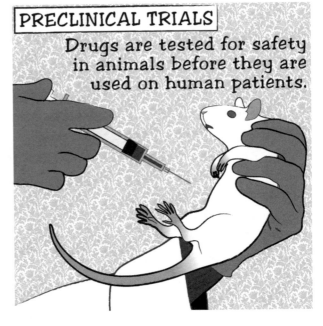

PRECLINICAL TRIALS

Drugs are tested for safety in animals before they are used on human patients.

Preclinical development— safety testing in animals— is typically considered the last phase of drug discovery. Companies are usually required to test a drug candidate in at least two different animal species, and record data on toxicology, pharmacokinetics, and pharmacodynamics. Preclinical development are governed by FDA Good Laboratory Practices (GLP).

After promising preclinical results, a company makes an application to the FDA for **Investigational New Drug (IND)** status. Technically, the IND is an exemption from the legal requirement that drugs have been approved for marketing before being shipped across state lines. In practice, IND status allows drugs to begin clinical testing.

The FDA requires three main documentations for IND review. An IND application must include results of animal studies showing drug pharmacology and toxicology; manufacturing information to show that the drug can be mass-produced consistently; and detailed clinical protocols describing how the clinical trials will be performed, specifying the physicians who will oversee the trials and verifying that they will follow informed consent procedures.

Tricky Terminology

Pharmacokinetics (PK) measures what the body does to a drug—i.e. how the drug is absorbed, distributed, broken down, and excreted. **Pharmacodynamics (PD)** describes what the drug does to the body—how increasing concentrations of the drug influence potential toxicity. Both parameters are important in establishing safety and appropriate dosage.

Clinical Trials

Clinical trials of new drugs in humans are designed to test the safety of the drug and to test the efficacy of the drug—whether it cures or lessens the effects of disease. As we learned, clinical trials are subdivided into three main phases. Each phase is governed by the FDA guidelines that govern Good Clinical Practices (GCP). Let's explore each phase further.

Phase I tests the safety of the drug in a small number of *healthy* volunteers. Under close supervision, volunteers receive escalating doses of drug until any side effects start to appear, at which point the escalation is stopped. By this means, researchers establish the **maximum tolerated dose (MTD)** for a drug, which becomes a benchmark for the rest of the trials. Some Phase I trials enroll patients instead of healthy volunteers for two main reasons. First, some drug candidates, such as oncology drugs, are too potentially toxic for administration to healthy individuals. Second, patients suffering from serious disease, for which there is no available treatment, may benefit so they are enrolled in Phase I trials. In either case, the number of participants in a Phase I study is relatively low, typically ranging between five and one hundred.

Tricky Terminology

Investigators conducting Phase I trials use one of two different statistical designs: **Single Ascending Doses (SAD)** and **Multiple Ascending Doses (MAD)**. SAD design means that small groups of subjects are given a single dose of drug. If no adverse effects are observed, a new group of subjects are given a higher dose, and so on, until the maximum tolerable dose is determined. MAD design means that small groups of subjects are given multiple low doses of the drug, with subsequent dose escalation for further groups based on safety data.

Phase II is the testing phase during which efficacy is first measured—in other words, does the drug act as it is intended to act in a patient population? Accordingly, all participants are patients. Researchers assess drug efficacy by monitoring established indicators of disease progression; for instance, tumor growth in cancer; glucose tolerance in diabetes; or blood pressure in hypertension. In this phase, researchers compare the drug's efficacy in the testing group to a control group of subjects who receive either a placebo, such as a sugar pill or a standard of care treatment—an approved treatment commonly used for the same indication. Both expected and unexpected side effects of drug treatment are closely monitored throughout

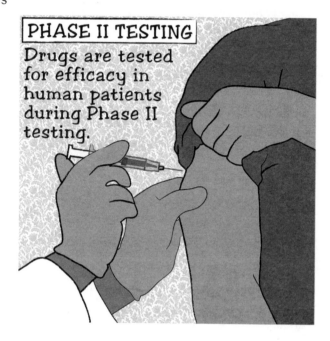

PHASE II TESTING

Drugs are tested for efficacy in human patients during Phase II testing.

Phase II, and indeed, the whole trial. Phase II trials include more participants than Phase I, typically ranging between 50 and 1000 participants; possibly including more than 1000, depending on the size of the patient population for which the sponsor is seeking approval.

GOING FURTHER

Scientists have found that placebos can have a significant impact on patient health. The **placebo effect** probably stems from a patient's expectation or belief that the drug will do them good. Because those patients receiving the actual test drug will, on average, have the same expectation (in addition to the drug effect), the placebo control is very important. Recent evidence actually shows that some patients suffering from chronic pain produce endogenous opioid painkillers in response to a placebo.

More rarely, the reverse of the placebo effect—the **nocebo effect**—can occur. In other words, patients have an expectation that a drug will not help them or that it will do them harm, and it does just that.

Phase III continues to test efficacy and safety, but in a much larger group of patients. A test group of 300 to 5,000 patients allows for a more statistically significant analysis of both efficacy and safety. Phase III trials, as well as Phase II, are usually randomized, double-blind studies, meaning that patients receive a random assignment to be in the group that receives the drug, the placebo, or the standard of care. Even the doctors involved in the trials do not know the patient group assignments, thereby ruling out any bias that may occur in assigning groups. If the investigational new drug appears to be effective, patients will be allowed to continue treatment after the trial's conclusion and before the drug's market approval.

If the drug meets the expected clinical endpoints during Phase III, the sponsoring company will submit a new drug application (NDA) for small molecule drugs, or a biologic licensing application (BLA) for biologics. Approval times vary, but typically the FDA will take between ten months and two years to review the data. After reviewing an application, the agency will either grant approval, deny approval, or request additional studies.

Even after a drug is approved, post-marketing safety monitoring continues in **Phase IV**. The purpose of Phase IV is to catch any unexpected adverse reactions of an approved drug. The FDA relies on patients, physicians, and drug companies themselves to report these incidents. Even though premarket clinical testing is extensive, unanticipated safety issues do occasionally show up, simply because the number of patients taking an approved drug is always many times larger than the number of individuals that the drug was tested on. If the side effects are serious, a drug may be pulled from the market, either voluntarily by the drug company or as required by regulatory agencies.

> ## Tricky Terminology
> **Pharmacovigilance** is defined by the World Health Organization as the science and activities relating to the detection, assessment, understanding, and prevention of adverse effects or any other drug related problem.

INDUSTRY NOTE

The diabetes drug troglitazone, which was marketed as Rezulin in the U.S., caused acute liver failure in some patients. Similarly, a cholesterol lowering statin called cerivastatin, which was marketed as Baycol in the U.S., caused at least 52 deaths from kidney failure. Both of these drugs were quickly pulled from the market.

Sometimes independent researchers analyze clinical trials after drug approval. This highly controversial analysis is known as **meta-analysis** and involves taking data from several clinical trials for the same drug. Some experts argue that such a purely statistical analysis is questionable since it does not take into account clinical trial experimental design. Nonetheless, meta-analysis has had a profound impact on drug marketing and safety awareness.

GOING FURTHER

Meta-analysis has called into question the safety of Avandia, a diabetes drug, implicating the drug in increased risk of heart attacks. Of 14,371 patients taking Avandia, 86 suffered a heart attack during the study period, compared to 72 patients out of 11,634 who took a standard of care drug like metformin, or a placebo. In this case, the increased risk of heart attack would seem quite small. Often, doctors and patients need to carefully weigh increased risk associated with a certain drug against its likely benefit.

Patents

In the United States, drug companies will typically apply for patent protection during the early phases of drug discovery. Under current law, patent protection lasts for twenty years from the time of award, so patent protection in the market is very dependent on the duration of clinical trials. Expedition of the clinical testing, therefore, becomes critical to the drug company. Even so, some drugs may win approval with just a few years of patent protection left.

In recognition of this, the FDA grants each new drug a period of exclusivity, which means that the holder of an approved new drug application receives limited protection from new competition in the marketplace for the innovation represented by its approved drug product. For small molecule drugs, the exclusivity period is five years, with a 6 month extension for products developed for pediatric use. Orphan drugs—products developed to treat a patient population of less than 200,000— receive a seven year exclusivity period. Biologic drugs are granted a twelve year exclusivity period. This longer exclusivity period for biologics is in recognition of the significantly greater manufacturing costs associated with biologics.

INDUSTRY NOTE

In the early 1980's the Orphan Drug Act provided incentives for researchers to develop drugs for treating rare diseases—those afflicting fewer than 200,000 people in the U.S.. Such diseases include multiple myeloma, Huntington's disease, Lou Gehrig's disease, cystic fibrosis, Parkinson's disease, and phenylketonuria. Because of the smaller potential market for orphan disease drugs, drug companies have traditionally focused research efforts on diseases such as cancer and heart disease. The Orphan Drug Act introduced tax breaks and an extra two years of market exclusivity, making orphan drug research a more attractive investment. Since 1983, about 250 orphan drugs have made it to market. Prior to the Orphan Drug Act, fewer than 10 such drugs received approval for marketing.

Generic: Small Molecule

The term generic refers to a small molecule drug that is an exact molecular copy of an innovator drug. If a generics manufacturer can demonstrate this molecular equivalence, than they may rely on the safety and efficacy data generated by the innovator company during human clinical trials in order to file what is referred to as an **Abbreviated New Drug Application (ANDA)**. An ANDA is filed after the patent protection and exclusivity periods have expired for the brand name drug. Because the drug **formulation** of a generic compound may be slightly different than that of the innovator, clinical data may be required in order to demonstrate that the generic drug has equivalent **bioavailability** to the original. If the ANDA is approved, the generic drug manufacturer is granted a 180-day exclusivity period, after which other generics manufacturers may commence manufacturing other generic versions. Because of this increased competition, the price of generic drugs is significantly reduced.

Tricky Terminology

A drug **formulation** refers to the different chemical substances, including the active drug, that are combined to produce a final medicinal product. Substances added may include stabilizing agents, bulking agents, and solubilizing agents. Any of these things may affect the active ingredient's **bioavailability.** Bioavailability refers to the total amount that reaches the target tissue and the time taken to get there after administration.

Biosimilar: Large Molecule

Because large molecule drugs are structurally so much more complex than small molecule drugs, it is impossible, using today's technology, to demonstrate that a large molecule drug produced by a company other than the innovator company is identical to the original product. For this reason, the term generic is not used when referring to copies of biologic drugs. A new term, biosimilar, has been introduced for this purpose.

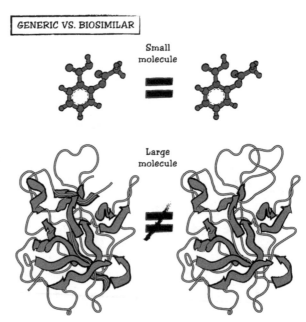

GENERIC VS. BIOSIMILAR

Small molecule

Large molecule

Biosimilars have been able to gain marketing approval in the EU since 2006. In 2009, the United States Congress passed the Biologics Price Competition and Innovation Act, creating the legal authority for the FDA to approve biosimilar drugs, but leaving the scientific guidelines regarding testing and approval up to the agency itself. In February 2012, the FDA released a draft guidance outlining what is likely to be required of companies hoping to enter the biosimilar space. Akin to the EMA guidelines, the FDA draft guidance states that makers of biosimilar products be required to present data from analytical studies to demonstrate the biosimilar product is *highly* similar to the reference product. The degree of likeness demonstrated in the analytical studies will dictate the extent to which animal and human testing be required. In most cases, it is expected that some degree of animal and human testing be considered necessary to adequately show safety and efficacy to the biosimilar product. Any reduction in preclinical or clinical trials will have to be scientifically justified by the biosimilar company. As with any drug, companies are expected to have in place vigorous post-market safety monitoring plans.

EXCERPT FROM BIOTECH PRIMER BLOG

First RNAi Drug to Show Promise in Humans

RNA interference, or RNAi, is a gene silencing technique that was discovered in the 1990's to occur naturally in petunia plants. Since its initial discovery, variations of the technique have been discovered in many different cell types, including mammalian cells. The basic idea is this: cells contain enzymes that recognize double stranded RNA and destroy it. This is in part because the presence of double stranded RNA could signify a viral infection, so its destruction may have evolved as a type of cellular defense against viruses. The petunia researchers also discovered that the production of double stranded RNA—termed in this case RNAi—is used by cells themselves at different points in their development to ensure life cycle appropriate expression of certain proteins—in the case of petunias, pigmentation proteins.

For the past decade, a number of different research groups have attempted to adapt this naturally occurring mechanism to drug development, by inducing the formation of double stranded RNAs that correspond to the genes whose expression the researcher is trying to block. Cellular enzymes—called Dicer and RISC—then degrade the RNA message. Without an RNA message, no protein is made.

This technique has worked beautifully in cells grown in the lab (tissue culture), as well as in some animal models. There has been less success, however, with demonstrating therapeutic results in humans. This is thought to be due largely to difficulties in getting the therapeutic RNAs to the right location within the human body, in high enough concentrations to be effective at gene silencing.

Clinical researchers at St. Jude's Children's Research Hospital in Memphis, TN, have achieved initial success of RNAi therapy used to prevent

Excerpt From Biotech Primer Blog Continued From Previous Page

respiratory syncytial virus infection (RSV) (Proceedings of the National Academy of Sciences, May 11, 2010, 107(10):8800). RSV is a leading cause of viral pneumonia in infants. Participants in the study—healthy adult volunteers—were treated with a nasal spray containing the RNAi therapeutic meant to interfere with the production of one of RSV's surface proteins, or a saline solution as a placebo, and then subsequently exposed to the virus (RSV does not make healthy adults seriously ill). Participants who received RNAi treatment prior to exposure to the virus were significantly less likely to have a measurable viral infection.

This is very good news, because it suggests that the use of RNA-based therapeutics may soon be a reality in human patients. Because these drugs promise to be highly specific—blocking the expression of a particular protein known to play a well-defined role in a given disease—they are likely to have very few adverse reactions, and to be highly effective. Delivery of this therapeutic to lung tissues was aided by the researchers' ability to use a nasal spray; delivery to other tissue types may continue to be a challenge. However, with advances in drug delivery technology and chemistry for making more stable RNA therapeutics, we can expect to see similar results for other diseases someday soon.

http://www.BiotechPrimerBlog.com

In this chapter, we have outlined the steps necessary for taking a drug from the lab to the market. In the next chapter, we will describe the large scale manufacturing of biologic drugs—biomanufacturing.

CHAPTER TEN
Putting The BIO In Biomanufacturing

Biomanufacturing is the making of biological products from living cells. Biomanufacturing plants produce a wide range of products, including enzymes, vaccines, therapeutic antibodies, and antibiotics. Manufacturing a biotechnology product for the marketplace involves producing very large quantities of the product, or large scale production. To reach the end stage of scale-up and manufacturing, the product must have gone through earlier production stages of product development, or the "**product pipeline**". These stages include identification of a drug candidate, all of the various steps in research and development, and product testing for safety and efficacy through clinical trials. For any biomanufacturing company, regulatory issues are a major consideration. Companies must follow the FDA's current Good Manufacturing Practice (cGMP) guidelines for scale-up and manufacturing process to ensure safety and purity of the product. In this chapter, we will examine some of the key components of the biomanufacturing process.

> ### Biomanufacturing companies must follow the FDA's cGMP guidelines.

Biomanufacturing Process Overview

As an example, we will focus on the process of producing an enzyme from living bacterial cells, though the process is also done with other cells, including insect, mammalian, and fungi. The choice of what cell type to use varies depending on the type of product being produced. Proteins with a relatively simple structure, such as human insulin, can be successfully produced in bacterial cells. Proteins with a more complex structure, such as monoclonal antibodies, need to be produced in correspondingly more complex mammalian cells. Fungal cells are often used to produce antibiotics such as penicillin.

The process of genetically engineering bacterial cells to make a specific protein begins with isolating the gene that encodes for that particular protein. Next recombinant DNA technology is used to insert the gene of interest into a bacterial plasmid creating a recombinant plasmid as described in Chapter Six. These recombinant plasmids then are transferred into bacterial cells, creating transformed cells. The biomanufacturer then cultures the recombinant bacterial cells, providing them with the proper environment and nutrients to grow and replicate. Once the transformed bacteria start growing, they will naturally make recombinant proteins from the new gene. This initial culture will provide a small scale batch of cells containing the newly inserted gene of interest.

The next steps are the scale-up processes, which first employs a bench top **bioreactor** (about ten-liter capacity) and then a pilot scale bioreactor (about 100 liter capacity). A bioreactor is a large vessel that grows bacterial cultures in great volumes by maintaining all the proper growing conditions, such as appropriate nutrient concentrations, pH levels, temperature, and oxygen levels. These growth conditions are closely monitored and adjusted where appropriate. After

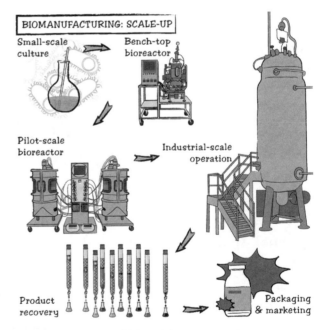

BIOMANUFACTURING: SCALE-UP

Small-scale culture

Bench-top bioreactor

Pilot-scale bioreactor

Industrial-scale operation

Product recovery

Packaging & marketing

successful production in the pilot scale reactor, the process moves into industrial scale operations that use much larger bioreactors—sometimes as large as one hundred thousand liters.

Once the cells in a large scale bioreactor reach their maximum density, the product is recovered. In some cases, the cells have secreted the protein into the growth media, but in many other cases, the cells must be harvested and broken open—lysed—in order to release the protein product. In either case, the protein product must be purified away from other cellular molecules before being packaged as a therapeutic.

GOING FURTHER

We mentioned earlier that Chinese hamster ovary cells (CHO cells) and mouse myeloma non-secreting cells (NSO cells, pronounced "NS zero") are common favorites for mammalian cell culture. This also applies for the production of biopharmaceuticals, especially monoclonal antibodies. These cell types are well established, understood, able to be grown in large volumes, and produce high yields of recombinant proteins. Both CHO and NSO cells process proteins in a manner very similar to that of human cells. In other words, proteins produced in these cells are likely to be folded in the same way as in a human cell. CHO and NSO cells give rise to similar or identical post-translational modifications—such as the addition of sugar groups, known as glycosylation. Both cell lines have **generally regarded as safe (GRAS)** status by the FDA for therapeutic protein production—in other words, drug companies can use them to manufacture product without first demonstrating their safety.

Critical Steps in a Production Campaign

The scale-up of a cell culture process can be difficult and time consuming. Because bacterial cells replicate quickly, scale-up of bacterial cultures may in some cases be completed in less than a week; scale-up of mammalian cells may take several weeks to more than a month before yielding a product. The entire process of producing a biotech product from start to finish is often called a "campaign" and is usually divided into two main parts: **upstream processing** and **downstream processing**. Upstream processing refers to the production of the protein product by using cells (microbial, insect, fungal, or mammalian) growing in culture. Downstream processing refers to the recovery, purification, formulation, and packaging of the protein product.

INDUSTRY NOTE

During the R&D phase, the company researchers develop initial production methods on a small scale. However, they must eventually determine a viable final physical formulation of the product, which could be a tablet, an aerosol inhaler, an injectable liquid, a patch, or a cream. Almost all biologic drugs currently available must be taken as an injectable or administered intravenously; small molecule drugs are more typically formulated as tablets. Using all of the R&D data from the earlier production steps, biomanufacturers have to figure out optimal production methods for the particular formulation of their product to produce enough for the intended market.

Cell Banks

Upstream processing begins with the cells engineered to make the protein product. Once they create the desired cell line, researchers **cryopreserve** the line by freezing a large number of vials of the cells in order to create a **cell bank**. Cell banking involves a two-tiered system: a **master cell bank (MCB)** and a **working cell bank (WCB)**. The working cell bank is created from one vial of cells from the master cell bank. Once established, the working cell bank is used to

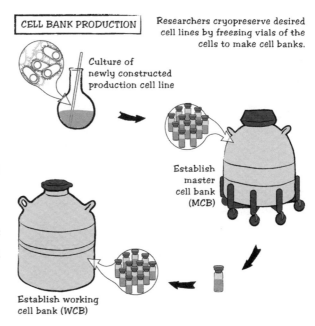

CELL BANK PRODUCTION

Researchers cryopreserve desired cell lines by freezing vials of the cells to make cell banks.

Culture of newly constructed production cell line

Establish master cell bank (MCB)

Establish working cell bank (WCB)

produce batches of product in the scale-up process, because using the same stock of cell line reduces the chance of mutations associated with serial cultures. The master cell bank functions as a reserve of cells that is only used when absolutely necessary. To protect the integrity of the cell lines and to take precautions against loss, companies usually maintain cell banks in two or more locations within the facility and in one location off-site.

Scale-Up and Manufacturing Process

A production campaign begins with the withdrawal of cells from the working cell bank. The cells are thawed, and a cell culture is initiated in a flask containing a small volume of growth media to provide the nutrients and optimum environment for cells to survive. By gradually transferring the growing cells into successively larger growth vessels, the biomanufacturer accomplishes scale-up. As long as the growth environment remains favorable, cells will continue to divide, and in doing so, produce increasing amounts of product. Towards the end of the scale-up process, cells are growing in production vessels that will typically hold fifteen to twenty thousand liters of a mammalian cell line, or as much as one hundred thousand liters of a bacterial cell line.

Scientists continue to test cell viability, product concentration, and product activity at each incremental step in the scale-up. They must also monitor the physical environment in which cell cultures are growing to optimize temperature, pH, nutrient concentration, and oxygen level. During the early scale-up steps, laboratory technicians perform this monitoring manually. The monitoring process becomes automated after the cell culture is large enough to be grown in bioreactors.

It is also critical during the scale-up and manufacturing stages to test the cultures for contamination by bacteria, yeast, or other microorganisms. Any contamination of a culture ruins the entire batch of product, costing the drug company millions of dollars and a great loss of time. There are very strict protocols for maintaining aseptic conditions at all times during the scale-up and manufacturing stages.

All of the monitoring crucial to the success of the scale-up and manufacturing stages of product development is the responsibility of the **Quality Control** (**QC**) and **Quality Assurance** (**QA**) departments within the drug company. The QC department conducts routine testing during the product development stages well before the product is at the stage of marketing, ensuring that the scale-up and manufacturing processes meet certain standards. The QA department is usually responsible for ensuring quality objectives are met and also handles reporting them as a product gets closer to marketing. Of course, QC and QA oversight procedures include adherence to FDA cGMP guidelines to ensure safety and purity of the product.

Harvesting and Purification Process

In the downstream phase of manufacturing, the protein product is isolated from the cells that produced it. Extraction of proteins from within the cell, or intracellular proteins, requires special protocols, which usually involve bursting the cells open to release the protein product. Scientists purify the therapeutic protein from the extract, separating it from the other cellular components. Proteins excreted from the cell, such as monoclonal antibodies, are easier to isolate because scientists can skip the extraction step.

The purification steps that separate the protein product from other cellular components use column chromatography methods, as described below. Purification of protein mixtures by **column chromatography** separates proteins based on the physical and chemical properties of size, shape, or electrical charge. Typically, the cell extract needs to be serially processed by a variety of different column types in order to achieve a pure product.

Column Chromatography

Chromatography allows scientists to separate individual molecules from a complex mixture based on size, structure, or electrical charge. To perform chromatographic separation, researchers pass the complex mixture through a solid **matrix**—the **stationary phase**—which separates different molecules.

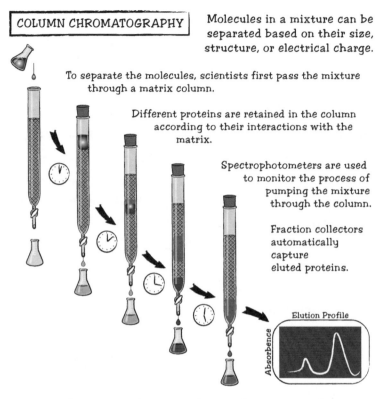

COLUMN CHROMATOGRAPHY

Molecules in a mixture can be separated based on their size, structure, or electrical charge.

To separate the molecules, scientists first pass the mixture through a matrix column.

Different proteins are retained in the column according to their interactions with the matrix.

Spectrophotometers are used to monitor the process of pumping the mixture through the column.

Fraction collectors automatically capture eluted proteins.

Elution Profile

Absorbence

Proteins are usually separated using columns that contain the matrix. The protein mixture is passed through the column under gravity and with the application of moderate pressure (**fast protein liquid chromatography** or **FPLC**) or high pressure (**high performance liquid chromatography** or **HPLC**) to increase speed and resolution. FPLC is generally the most effective and most commonly used method for proteins separation.

Proteins not retained by the column flow straight through, while proteins retained by the column follow later or are washed off, or **eluted**, under different conditions. Scientists monitor protein elution from the column using a **spectrophotometer**, which measures the ability of proteins to absorb UV light, providing an elution "profile". **Fractions** corresponding to the different elution times are collected automatically in a fraction collector.

Ion Exchange Chromatography

Ion exchange chromatography separates proteins based on their charge. Some amino acids, such as glutamate and aspartate, have a negative charge at neutral pH, while others, such as arginine and lysine, have a positive charge. Therefore, proteins can have an overall negative, positive, or neutral charge, depending on their amino acid composition. The ion exchange matrix used to separate them contains beads coated with a negatively or positively charged chemical.

In a chromatography matrix composed of positively charged beads (called **anion exchange**), for example, positive and neutral proteins flow straight through the matrix, while negative proteins are retained because the negatively charged proteins are attracted to the positively charged coated beads. Scientists elute the matrix bound protein by applying a gradient salt solution. The application of the salt solution causes **ions** to exchange with, and displace, proteins **adsorbed** to the beads. The displaced proteins then flow out of the bottom of the column. By gradually increasing the salt concentration in the solution, scientists make the protein molecules elute from the column according to the strength of their charges. Those molecules with a very strong ionic interaction require a higher salt concentration and elute later in the gradient. All of the material that elutes from the column is collected in a numbered series of test tubes, and these individual samples are called fractions.

Affinity Chromatography

Affinity chromatography works on a similar principle to ion exchange chromatography, but the matrix is designed to separate protein molecules according to their shape instead of their charge. Only proteins with a "pocket" complementary to the unique shape attached to the bead will bind to the matrix in the column. After the unbound proteins are washed away, the bound proteins can be eluted with a soluble form of the shaped chemical that replaces the bound form.

Size Exclusion Chromatography

Size exclusion chromatography, commonly called gel filtration, works by a different principle and separates proteins based on their size; larger proteins flow through the matrix faster than smaller ones. The matrix contains porous beads with pores of a defined size. Large proteins cannot enter the beads and simply flow around them. Smaller proteins can enter the beads and pass through them, resulting in a longer time taken to reach the bottom. Proteins of intermediate size either do or do not flow through the beads. The result is that different sized proteins gradually separate from one another and are collected in different column fractions once they pass through the column. For gel filtration columns to work effectively, they must be much longer and thinner than other types of columns to enable resolution of different size proteins.

Formulation, Fill, and Finish

After purification, the protein product is ready to be formulated according to the established R&D specifications that take into account product stability and delivery method. The company may wish to use product **excipients**, which are pharmacologically inactive ingredients that enhance the drug product. Such enhancements may include color additives, time release factors, and bulking agents such as mannitol. Stability agents such as antioxidants, buffers, and surfactants (to decrease clumping) may also be added.

Finally, the company establishes product fill concentration, and determines the labeling and packaging for large scale distribution, completing the biomanufacturing process.

BioSimilars vs. Generic Drugs: What's In a Word?

"Biosimilar" is a buzzword that isn't going away anytime soon. Sometimes also referred to as "follow-on biologics," biosimilar is the term used for a biologic drug that is produced using a different cell line, master cell bank, and/or different process than the one that originally produced the product.

How do biosimilars differ from so-called generic drugs? Generic drugs are essentially "copycat" versions of small molecule drugs—drugs that can be synthesized in the lab by following standardized, predefined procedures. Using well established analytic techniques, the generic version of a small molecule drug can be demonstrated to be chemically and structurally identical to the innovator drug. Biologic drugs, however, are much more complex than small molecule drugs. These drugs are proteins that must be produced by a living cell—they cannot be chemically synthesized in the lab by following a standard set of procedures. The cell makes these proteins by following a recipe provided by a short sequence of DNA—a gene—that is inserted into the cell.

Here's the catch: even if two different cells are provided the exact same recipe, the final product may be slightly different. This may happen even if the two cells are of the same type—very slight environmental differences can have a profound effect on how a given cell follows a particular recipe. This makes intuitive sense—we know that we can follow the same recipe that the gourmet chef at our favorite restaurant follows, but somehow, the dish doesn't taste quite as good. Slight differences such as the exact temperature of the frying pan, the brand of the heavy cream, and even the humidity level and elevation of the kitchen will influence the final outcome. The restaurant industry depends on these "trade secrets" to keep us coming back for more.

Excerpt From Biotech Primer Blog Continued From Previous Page

Complicating matters further for biosimilar products is the fact that because biologic drugs are structurally much more complex than their small molecule counterparts, it is not currently possible to demonstrate conclusively that a biosimilar drug is in fact identical to the original biologic drug. Thus, since we know that there is a high likelihood that a biosimilar drug is not identical to the original biologic, and we have no way of precisely measuring whatever differences may exist, the term "biosimilar" is used rather than "generic", which implies identity.

http://www.BiotechPrimerBlog.com

In this chapter, we have reviewed the critical steps of a biomanufacturing campaign, including cell bank creation, cell culture scale-up, and product purification.

Conclusion

In the preceding chapters, we have taken an in-depth look at the biotech industry, and in particular, the science that drives it. From cell structure to protein structure; gene expression to genetic variation and genetic engineering; the human immune response to the production of antibodies for biotech application; and finally drug discovery, drug development, and biomanufacturing—we have discussed the key concepts and technologies that impact current biotechnology developments.

It is our hope that you will use this book continuously as a reference to support your growth as a biotechnology professional. Although the industry itself is constantly changing, these fundamental concepts upon which it is built will remain important for years to come—and decision-makers who understand these fundamentals will be better able to evaluate and predict new trends.

> Although the industry itself is constantly changing, these fundamental concepts upon which it is built will remain important for years to come...

More than anything else, we hope that your understanding of the science behind biotechnology will serve to increase your enthusiasm for this exciting and truly life-changing industry.

Glossary

A

Active Site: The portion of an enzyme that attaches to the substrate.

Adaptive Immune Response: The response of antigen-specific lymphocytes to antigen.

Adenine: One of the four nucleotide bases that make up DNA.

Adenosine Triphosphate (ATP): A compound used by cells to store energy and to fuel metabolic processes.

Adsorbed: Attracted to and maintained on the surface of a solid surface, as in chromatography.

Affinity Chromatography: A chromatographic method that makes use of the specific binding of one molecule for another.

Allele: One of a number of different forms of a gene. Each person inherits two alleles for each gene, one allele from each parent. These alleles may be the same or may be different from one another.

Amino Acid: One of twenty different molecules that combine to form proteins. The sequence of amino acids in a protein determines the protein's structure and function.

Angiogenesis: Growth of a network of blood vessels that penetrates into cancerous growths, supplying nutrients and oxygen and removing waste products.

Anion Exchange Chromatography: A type of column chromatography in which positively charged proteins are captured by a negatively charged chromatography matrix.

Antibody: A protein produced by the immune system that binds to a specific antigen.

Anticodon: A specific three nucleotide sequence in transfer RNA that is complementary to a codon (a three nucleotide sequence in messenger RNA) that specifies an amino acid in protein synthesis. When a codon and anticodon bind (because they are complementary strands) the amino acid attached to the transfer RNA is able to connect to the growing amino acid strand, which eventually forms a protein.

Antigen: A foreign substance which, when introduced into the body, stimulates an immune response.

Antigen Presenting Cell: A highly specialized cell that is able to process antigens and display their peptide fragments on the cell surface together with molecules required for lymphocyte activation.

Antiserum: The fluid component of clotted blood from an immune individual that contains antibodies against the molecule used for immunization.

Apoptosis: The process of cell self-destruction.

Assay: A test.

Atom: A particle, made up of a nucleus and one or more orbiting electrons, which is the basic unit of a chemical element.

B

B-Cell: An antibody producing cell of the immune system.

B-Cell Receptor: The cell-surface receptor of B-cells for specific antigen.

Bacterium: A group of small, single-celled, prokaryotic microorganisms.

Base: One of the molecules—adenine, guanine, cytosine, thymine, or uracil—which form part of the structure of DNA and RNA molecules. The order of bases in a DNA molecule determines the structure of proteins encoded by that DNA. See Nucleotide.

Base Pair (bp): Two complementary nucleotide bases joined together by chemical bonds. The base adenine pairs with thymine, and guanine pairs with cytosine.

Basophil: White blood cells that participate in the inflammatory response.

Bioactive: Having an effect on a biological system.

Bioavailability: Measure of the rate and the total amount of drug that reaches the target tissue after administration.

Biologics: Products of living organisms or cells used in the treatment or management of a disease.

Biologics Licensing Application (BLA): An application for marketing approval for a biologic drug; submitted to the FDA upon successful completion of Phase III clinical trials.

Biomanufacturing: The use of living cells to produce a biological product. An example is a therapeutic protein.

Biomarker: A physiological event or molecule that can be measured. Examples include the presence or absence of a protein or a mutated gene. Biomarkers are often used to indicate the presence or progression of a disease.

Biopharmaceutical: Drugs, either chemical compounds or biologics, made using the method of rational drug design.

Bioreactor: The usually stainless steel tank used to grow living cells that produce biologics. A bioreactor can range in size from a few liters up to 100,000 liters.

Bioremediation: A technique used to remove or neutralize contaminants in a particular environment using living organisms.

Biosciences: A term meant to encompass both biotech and pharmaceutical companies.

Biosimilar: A biologic drug that is produced using a different cell line, master cell bank, and/or different process than the one that originally produced the product.

Biotechnology: The use of cellular and biomolecular processes to solve problems and make useful products.

C

Carriers: A person who has inherited a genetic trait or mutation, but who does not display that trait or show symptoms of the disease.

CCR5 Gene: C-C chemokine receptor type 5 gene codes for a protein on the surface of white blood cells that is involved in the immune system as it acts as a receptor for chemokines. Many forms of HIV, the virus that causes AIDS, initially use CCR5 to enter and infect host cells.

Cell: The basic subunit of any living organism, typically containing at a minimum genetic material, an energy-producing system, and protein-making machinery, all surrounded by a membrane.

Cell Bank: A uniform population of cells, stored under defined conditions, typically frozen at -80 degrees Celsius or colder. The assumption is that each vial of cells is comparable, and may be used in a consistent manner after being thawed. See Master Cell Bank and Working Cell Bank.

Cell Line: The descendants of an original group of cells taken from an organism.

Cell Lysate: The cellular debris and fluid produced by brakeing open (lysing) a cell.

Cell Membrane: Barrier made out of fats and proteins that separates the inside of the cell from the outside.

Cell Wall: A stiff covering around the plasma membrane of certain non-animal cells, such as plants and many types of bacteria.

Channel Protein: A protein that spans the cell membrane, allowing substances to pass through from the outside to the inside of the cell.

Chemokine: Signaling molecules that are involved in the activation and migration of immune system cells. Chemokine signaling plays a key role in the inflammatory response.

Chimeric Antibody: Antibodies whose gene sequence consists of DNA from two different species. Typically, the term refers to antibodies whose DNA is between 10% and 25% mouse origin, with the remaining sequence being of human origin.

Chromosome: A long strand of DNA found within cells. Chromosomes contain both genes and regions of DNA that do not code for proteins.

Codon: A set of three nucleotide bases in a DNA or RNA sequence, which together code for a unique amino acid. For example, AUG (adenine, uracil, guanine) codes for the amino acid methionine.

Column Chromatography: A laboratory technique for the separation of a mixture of proteins based on each protein's unique chemical and physical characteristics.

Combination: A product comprised of two or more FDA-regulated categories.

Combinatorial Chemistry: A technique for rapidly and systematically assembling different combinations of molecules to create tens of thousands of diverse compounds. Used in drug discovery screening assays to identify potential useful therapeutic candidates.

Companion Diagnostic: A diagnostic used by a physician to inform his prescribing decision.

Contract Research Organization (CRO): A company that conducts preclinical or clinical trials for another company on a contract basis.

Current Good Manufacturing Practices (cGMP): FDA guidelines governing the manufacturing of biopharmaceuticals.

Cytochrome P450 (CYP): A family of enzymes important for drug metabolism.

Cryopreserve: The preservation of cells or whole tissues by cooling to low sub-zero temperatures.

Cytokine: Proteins made by cells that affect the behavior of other cells.

Cytoskeleton: Network of protein filaments just underneath the cell membrane that gives the cell shape.

Cytosine: One of the four nucleotide bases that make up DNA.

Cytotoxic T-Cells: T-cells that can kill other cells, typically virus-infected cells or tumor cells.

Cytotoxins: Proteins made by cytotoxic T-cells that participate in the destruction of other cells.

D

Deletion Mutation: One or more nucleotides is removed from a DNA sequence during the replication process.

Denature: Dramatic change in the conformation of a protein, usually unfolding caused by heat or exposure to chemicals such as acids. Most often will result in loss of biological function.

De Novo: From the beginning.

Deoxyribonucleic Acid (DNA): The molecule that encodes genetic information. DNA is a double-stranded helix held together by bonds between pairs of nucleotides.

Deoxyribose: A type of sugar which is a component of DNA.

Diagnostic (Dx): A test used to identify a disease or disorder, or to monitor the progression of treatment. Routine diagnostics are broad screening tools, whereas a specialty diagnostic screens for a specific disease.

Double Helix: The shape of the DNA molecule.

Downstream Processing: The phase of a biomanufacturing campaign that consists of harvesting, purifying, and formulating the product.

DNA Helicase: The enzyme that separates the two strands of a DNA molecule prior to replication.

DNA Ligase: An enzyme that acts as a molecular glue, gluing two pieces of DNA together.

DNA Polymerase: The enzyme that replicates DNA.

Domain: Portion of a protein that has its own structure.

Drug Candidate: A small molecule or biologic that is being tested for its therapeutic potential.

Drug Development: The process of testing therapeutic molecules for safety and efficacy in animals and humans, and developing appropriate formulation, delivery, and manufacturing methods.

Drug Discovery: The process of identifying molecules with a therapeutic effect against a target disease.

Drug Target: Organ, tissue, or molecule involved in a disease that is modified or affected by a potential therapeutic.

E

EC50: Drug concentration at which one-half of the maximum desired effect is observed.

E. coli: Common bacterium that has been studied intensively by geneticists because of its small genome size, normal lack of pathogenicity, and ease of growth in the laboratory.

Efficacy: The ability of a substance to produce a desired clinical effect.

Electroporation: The act of applying an external electrical field to a cell membrane in order to increase its permeability. The technique is used as a way of introducing DNA into bacterial cells.

Electrophoresis: A method of separating large molecules (such as DNA fragments or proteins) from a mixture of similar molecules. An electric current is passed through a medium containing the mixture, and each kind of molecule travels through the medium at a different rate, depending on its electrical charge and size.

Elute: To wash off or remove adsorbed material from a solid support.

Enzyme: A protein that enables a biochemical reaction in a cell.

Eosinophil: White blood cells thought to be important chiefly in the defense against parasitic infections.

Epitope: The structural feature of an antigen molecule to which an antibody binds.

Epidermal Growth Factor (EGF): A protein which binds to the epidermal growth factor receptor and stimulates cell division.

Epidermal growth factor receptor (EGFR): Transmits proliferation signals to many different cell types. Mutations in this protein are associated with different types of cancer.

Erythrocyte: Red blood cells.

Eukaryote: Cell or organism with membrane-bound, structurally discrete nucleus and other well developed subcellular compartments.

Evolution: The natural selection of beneficial changes.

Ex vivo: Pertaining to a biological process or reaction taking place outside of a living organism.

Excipient: Anything present in a drug product that is not the active ingredient. Includes components such as bulking and stabilizing agents, preservatives, salts, solvents, and water.

Extremophiles: Microbes that thrive in extreme environments.

F

Fast Protein Liquid Chromatography (FPLC): The process of separating proteins using columns with an application of moderate pressure.

Formulation: The process by which different chemical substances, including the active drug, are combined to produce a final medicinal product.

Fractions: A separate portion of a mixture, often used to describe the part that contains a particular molecular species.

G

G Protein Coupled Receptor (GPCR): A type of receptor protein that is targeted in over 25% of all biotech drugs.

Gamete: Reproductive cells; the ovum or sperm.

Gene: A length of DNA that codes for a particular protein.

Gene Expression: The process by which the information in a gene is used to create proteins.

Gene Product: The protein produced by a gene.

Generally Regarded As Safe (GRAS): A special status afforded by the FDA to ingredients and methods that have a proven, longstanding history of causing no harm to humans or animals.

Genetic Code: Set of rules by which the information encoded in DNA is translated into proteins.

Genetic Engineering: Altering the genetic material of cells or organisms in order to make them capable of producing new substances or performing new functions.

Genetic Variation: Differences in DNA sequence that occurs between individuals.

Genetically Engineered Organism (GEO): An organism whose DNA has been altered using genetic engineering techniques.

Genetically Enhanced Organism: An organism whose DNA has been altered using genetic engineering techniques.

Genetically Modified Organism (GMO): An organism whose DNA has been altered using genetic engineering techniques.

Genome: All of the genetic material in the chromosomes of a particular organism.

Germ Cells: Reproductive cells; the ovum or sperm.

Glycosylation: Adding one or more carbohydrate molecules onto a protein after it has been built by the ribosome.

Golgi Body: Cellular organelle which sorts and sends proteins to their appropriate location within the cell.

Granulocyte: A type of white blood cells characterized by the presence of small particles in their cytoplasm.

Growth Medium: A liquid or gel designed to support the growth of cells or microorganisms. There are different types of media for growing different types of cells.

Guanine: One of the four nucleotide bases that makes up DNA.

H

Half-Life: The amount of time it takes for 50% of a drug given to a patient to be eliminated or destroyed by natural processes.

Haplotype: Set of single nucleotide polymorphisms (SNPs) on a chromosome which are statistically associated.

International HapMap Project: An international project that seeks to determine any disease associations with specific haplotypes.

Hematopoiesis: The generation of the cellular components of blood, including red blood cells, white blood cells, and platelets.

Hematopoietic Stem Cell: Stem cells found in the bone marrow that have the potential to develop into any of the different types of blood cells found in the body.

Hemostasis: Clotting of blood.

High Pressure Liquid Chromatography (HPLC): The process of separating proteins using columns with an application of high pressure.

High Throughput Screening (HTS): Automated trial-and-error testing, typically using robotics, of very large sets of chemicals or materials.

Homeostasis: The property of a biological system that regulates its internal environment and tends to maintain a stable, constant condition of properties such as temperature or pH.

Human Anti-Mouse Antibody (HAMA): Antibodies generated by the human immune system when a human is injected with an antibody produced by a mouse.

Human Epidermal Growth Factor Receptor 2 (HER2): HER2 is a receptor found on the outside of cells that when activated causes the cell to multiply. HER2 is over-expressed in 25% of all breast cancers.

Humanized Antibody: An antibody produced in a non-human species whose DNA sequence has been altered to make it more closely resemble a human antibody.

Hybridoma: Hybrid cell lines formed by fusing a specific antibody producing B-cell with a myeloma cell line.

I

Inflammation: A non-specific immune defense by the body in response to injury or the presence of foreign particles.

Innate Immune Response: The early phase of the host response to infection in which a variety of mechanisms recognize and respond to a pathogen. Innate immunity is present in all individuals at all times, does not increase with exposure to a given pathogen, and does not discriminate between pathogens. Inflammation is an example of an innate immune response.

Interferon: Cytokines that can induce cells to resist viral infection.

Interleukin: A generic term for cytokines produced by white blood cells. Specific interleukins may have either an inhibitory or a stimulatory effect on the immune system.

Ions: A positively or negatively charged atom.

Ion Channel: A channel protein through which ions are allowed to pass.

Ion Exchange Chromotography: A type of chromatography that depends on adsorbing a charged protein to a matrix carrying the opposite charge.

In Silico: Studies done on a computer; for example, modeling the structure and function of a protein.

In Vitro: Pertaining to a biochemical process or reaction taking place in a test tube as opposed to taking place in an organism.

In Vivo: Pertaining to a biological process or reaction taking place in a living organism.

Insertion Mutation: One or more nucleotides is added to the DNA sequence.

Investigational New Drug (IND): A drug which has gained FDA approval to be shipped across state lines, typically for clinical trials, but has not yet gained approval for marketing.

K

Kinase: An enzyme that transfers a phosphate group from ATP to a protein. Often, this results in activation of the recipient protein.

L

Large Molecule Drug: Another name for protein therapeutics. Large molecule drugs are too large to enter cells.

Leukocyte: General term for white blood cell.

Life Sciences: A term meant to encompass both biotech and pharmaceutical companies.

Ligand: Any molecule that binds to a specific site on a protein or other molecule.

Ligation: The act of "gluing" two pieces of DNA together, using the enzyme DNA ligase.

Liposome: Artificial lipid bilayer vesicle.

Lymphocyte: A white blood cell that is important in the body's immune response.

Lymph System: The system of lymphoid channels that drains extracellular fluid from the periphery.

M

Macrophage: A type of white blood cell that destroys pathogens by engulfing them.

Master Cell Bank: A culture of fully characterized cells distributed into separate vials, processed together in such a manner as to ensure uniformity. The master cell bank is usually stored at -80 degrees Celsius or colder (liquid nitrogen), and at two geographically distinct locations.

Matrix: In column chromatography, the solid phase to which proteins are adsorbed.

Maximum Tolerated Dose (MTD): Highest dose of a pharmacological treatment that will produce the desired effect without unacceptable toxicity.

Mechanism-Based Drug Design: The development of new therapeutics based on an understanding of the underlying disease mechanism.

Medical Device: An instrument, apparatus, implant, in vitro reagent, or other similar or related article, which is intended for use in the diagnosis of disease or other conditions, or in the cure, mitigation, treatment, or prevention of disease, or intended to affect the structure or any function of the body and which does not achieve any of its primary intended purposes through chemical action within or on the body.

Memory B-Cell: A B-cell that remembers the same pathogen for faster antibody production in future infections.

Messenger RNA (mRNA): The DNA of a gene is copied into mRNA molecules, which then serve as a template for the synthesis of proteins.

Meta-Analysis: Methods focused on contrasting and combining results from different studies, in the hope of identifying patterns among study results, sources of disagreement among those results, or other interesting relationships that may come to light in the context of multiple studies.

Mitochondrion: An organelle within a cell that generates most of the cell's energy.

Molecule: Two or more atoms connected by chemical bonds.

Monoclonal Antibody (mAB): An antibody produced by a single clone of cells, which therefore consistently binds to the same epitope of an antigen.

Monocyte: Precursor to macrophages.

Monogenic Disease: A disease that can be linked to a mutation in one specific gene.

Multicellular: Organisms made up of more than on cell.

Multiple Ascending Doses (MAD): A method for determining the maximum dose of a drug to be tested that consists of giving small groups of subjects multiple low doses of the drug, followed by subsequent escalation of the dose for further groups based on safety data.

Mutation: A change, deletion, or rearrangement in the DNA sequence that may lead to the synthesis of an altered or inactive protein or the loss of the ability to produce the protein.

Myeloma: A cancer of plasma cells.

N

Natural Selection: A mechanism of evolution whereby members of a population with the most successful adaptations to their environment are most likely to survive and reproduce.

New Drug Application (NDA): An application for marketing approval for a small molecule drug; submitted to the FDA upon successful completion of Phase III clinical trials.

Nuclear Magnetic Resonance (NMR): A technique used to determine protein structure by measuring the absorption of electromagnetic radiation at a specific frequency.

Nuclear Membrane: The membrane surrounding the nucleus.

Nucleic Acid: One of the family of molecules which includes the DNA and RNA molecules.

Nucleotide: The "building block" of nucleic acids, such as DNA and RNA molecules. A nucleotide consists of one of five bases—

adenine, guanine, cytosine, thymine, or uracil—attached to a sugar-phosphate group.

Nucleus: The membrane bound structure containing a cell's DNA found within all eukaryotic cells.

Nerve Growth Factor: A growth factor that binds to receptors on the surface of nerve cells, stimulating their growth.

Neutrophil: A major class of white blood cells. Play an important role in engulfing and killing foreign invaders.

Nocebo Effect: Patients have the expectation that a drug will not help them or that it will do them harm, and it does just that.

O

Organelle: Membrane-bound structures in a cell that have specialized functions, such as mitochondria and the nucleus.

Orphan Drug: A drug developed for a condition that affects fewer than 200,000 individuals in the U.S.

Osteoblast: A cell responsible for bone formation.

P

p53: A tumor suppressor gene. Mutations in p53 are linked to many different types of cancer.

pH: A measure of the acidity (hydrogen ion concentration) of a solution.

Pathogens: Microorganisms that can cause disease when they infect a host.

Peptide Bond: A type of chemical bond connecting amino acids.

Phagocytes: Cells capable of engulfing other cells.

Pharmaceutical: Any chemical substance intended for use in the medical diagnosis, cure, treatment, or prevention of disease.

Pharmacodynamics (PD): The study of the effect of a drug on the body; in particular, the effect of the drug as it relates to increasing dose.

Pharmacogenomics: The science of understanding the correlation between an individual patient's genetic make-up (genotype) and their response to drug treatment. Some drugs work well in some patient populations and not as well in others. Studying the genetic basis of patient response to therapeutics allows drug developers to more effectively design therapeutic treatments.

Pharmacokinetics (PK): The study of drug absorption, drug distribution within the body, drug metabolism, and drug excretion.

Phase I Clinical Trial: Initial studies to determine the metabolism and pharmacologic actions of drugs in humans, the side effects associated with increasing doses, and to gain early evidence of effectiveness; may include healthy participants and/or patients.

Phase II Clinical Trial: Controlled clinical studies conducted to evaluate the effectiveness of the drug for a particular indication or indications in patients with the disease or condition under study and to determine the common short-term side effects and risks.

Phase III Clinical Trial: Expanded trials after preliminary evidence suggesting effectiveness of the drug has been obtained, and are intended to gather additional information to evaluate the overall benefit-risk relationship of the drug and provide an adequate basis for physician labeling.

Phosphate: A chemical group consisting of an atom of phosphorous chemically bonded to four oxygen atoms.

Placebo: A fake treatment administered to the control group in a controlled clinical trial.

Plasma Cell: An antibody producing B-cell.

Plasmid: A small, circular piece of DNA that is separate from the cell's genome. Plasmids are manipulated in the laboratory to deliver specific genetic sequences into a cell.

Platelets: Blood cells that are required for blood clotting.

Point Mutation: A single nucleotide change in a DNA sequence.

Polyclonal Antibody: A mixture of antibodies that recognize different epitopes on the same antigen; each antibody is produced by a different B-cell.

Polygenic Disease: A disease that results from interactions among two or more genes.

Polymorphism: Small differences that occur between genomes.

Post-Translational Modifications: Biochemical changes made to a protein by the cell after it is made.

Preclinical Studies: The testing of experimental drugs in the test tube or in animals—the testing that occurs before trials in humans may be carried out.

Primary Response: The adaptive immune response after initial exposure to an antigen.

Product Pipeline: The series of products developed and sold by a company, ideally in different stages of their life cycle.

Progenitor Cells: The more differentiated offspring of stem cells that give rise to distinct subsets of mature blood cells.

Prokaryote: Cell or organism lacking a membrane-bound, structurally discrete nucleus and other subcellular compartments. Bacteria are prokaryotes.

Promoter: A segment of DNA located in front of a gene, which provides a site where an enzyme can bind to the DNA molecule, to initiate transcription.

Prostaglandins: Signaling molecules that have a variety of physiological effects, including smooth muscle contraction, platelet aggregation, and control of cell growth.

Protease: An enzyme that degrades proteins.

Protein: A biological molecule that consists of many amino acids linked together by peptide bonds. As the chain of amino acids is being synthesized, it is also folded into higher order structures. Proteins are required for the structure, function, and regulation of cells, tissues, and organs in the body.

PTEN: A human tumor suppressor protein. As such, it regulates cell growth and division.

Q

Quality Assurance (QA): The quality systems and processes used to control every step of pharmaceutical manufacturing to ensure that the product meets all of its specifications and quality attributes, and that all steps were done and documented in compliance with cGMP.

Quality Control (QC): The system of testing that confirms and measures the quality of raw materials, process intermediates, final product, and environmental samples.

R

R Group: In amino acids, the chemical group that varies and gives each amino acid its unique chemical and physical properties.

Rational Drug Discovery: The development of new therapeutics based on an understanding of the underlying disease mechanism.

Receptor: A protein usually found on the surface of a cell that binds to a specific chemical messenger, such as a neurotransmitter or hormone.

Recombinant DNA (rDNA): DNA molecules that have been created by combining DNA from more than one source.

Research Support Company: A company who does not actively engage in drug discovery, but supplies all of the tools and technologies required.

Restriction Enzyme (RE): A protein that recognizes specific, short nucleotide sequences and cuts DNA at those sites.

Resolution: The dissipation of an inflammatory response.

Ribonucleic Acid (RNA): A nucleic acid similar to DNA but based on the sugar ribose and containing the nucleotides guanine, adenine, uracil, and cytosine and typically single stranded.

Ribosome: The cell structures within which protein synthesis occurs.

RNA Interference (RNAi): A technique used to block the expression of a particular protein.

S

Scale-Up: The process of slowly increasing the volume of a cell culture from a few milliliters to several thousand liters.

Secondary Response: The antibody response induced by a second exposure to the antigen.

Sequence: The order of nucleotides in a DNA or RNA molecule, or the order of amino acids in a protein.

Single Ascending Dose (SAD): A method for determining the maximum dose of a drug to be tested that consists of giving small groups of subjects a single dose of the drug; if no adverse effects are observed, a new group of subjects is given a higher dose.

Signal Transduction: The general process by which cells receive information from their environment.

Single Nucleotide Polymorphism (SNP): A difference in one base pair between two DNA sequences.

Small Molecule Drug: A drug that is chemically synthesized in the lab. Small molecule drugs are small enough to enter cells.

Somatic Cell: A cell that is not an egg or a sperm cell.

Specrophotometer: A machine used to measure the transmittance of ultraviolet light through a solution in order to determine the concentration of protein in that solution.

Stationary Phase: In column chromatography, the column matrix.

Start Codon: The three nucleotide sequence that signifies the ribosome to start translating an mRNA sequence into a protein.

Stop Codon: The three nucleotide sequence that signifies the ribosome to stop translating an mRNA sequence into a protein.

Stromal Cells: Connective tissue in the bone marrow and other tissue types.

Substitution Mutation: A single nucleotide is exchanged for a different nucleotide in a DNA sequence.

Substrate: A molecule on which enzymes act.

Susceptibility Gene: A gene for which certain alleles correspond to a higher risk of developing a particular disease.

T

T-Cell: An immune system cell that recognizes specific pathogens based on the shape of its cell surface receptor.

T-Cell Receptor: Protein on the surface of T-cells to which antigen binds, activating T-cell.

T-Helper Cell: T-cells that help to fully activate antibody secreting plasma cells by secreting activating cytokines.

Tag SNP: A specific SNP that is used to identify a particular haplotype.

Target Validation: Determining if targeting a particular molecule thought to be involved in a disease mechanism will be a safe and effective means of therapy.

Template: The strand of DNA that is being used by the DNA polymerase to construct a new DNA molecule.

Tetramer: A protein composed of four subunits.

Therapeutic Window: Concentration range within which a drug is both effective and safe.

Thrombocytes: Cells required for blood clotting.

Thymine: One of the four nucleotide bases that makes up DNA.

Tissue Culture: The growth and maintenance of cells outside of a living organism in the lab.

Transcription: The process during which the information in a length of DNA is used to construct an mRNA molecule.

Transduce: To convert a chemical signal from one form to another.

Transfection: The process of introducing DNA into eukaryotic cells.

Transfer RNA (tRNA): RNA molecules that bind to amino acids and transfer them to ribosomes, where protein synthesis is completed.

Transformation: A process by which the genetic material carried by an individual cell is altered by incorporation of external DNA into its genome.

Transformed: Bacterial cells that have had foreign DNA introduced.

Transgenic Organism: An organism whose genome has been altered by the incorporation of foreign DNA.

Translation: The process during which the information in mRNA molecules is used to construct proteins.

U – Z

Unicellular: Organisms made up of only one cell.

Uracil: One of the four nucleotide building blocks of RNA.

Upstream Processing: The phase of biomanufacturing that consists of establishing cell banks and seeding and scaling up cell cultures.

Vasodilation: Dilation of a blood vessel.

Vector: A vehicle for the transfer of DNA from one organism to another. A plasmid is a common type of vector.

Working Cell Bank: A cell bank that is established from one of the master cell bank vials.

Xenotransplant: Transplantation of tissue or organs between organisms of different species, genus or family. A common example is the use of pig heart valves in humans.

X-Ray: A form of electromagnetic radiation.

X-Ray Crystallography: Technique for determining the three-dimensional arrangement of atoms in a molecule based on the diffraction pattern of X-rays passing through a crystal of the molecule.